# New Materials and Devices for Thermoelectric Power Generation

*Edited by Basel I. Abed Ismail*

Published in London, United Kingdom

New Materials and Devices for Thermoelectric Power Generation
http://dx.doi.org/10.5772/intechopen.110935
Edited by Basel I. Abed Ismail

Contributors
Basel I. Abed Ismail, Jehad H. Ismail Abed, Baljit Singh Bhathal Singh, Ebrar Yildirim, Övgü Ceyda Yelgel, Sikander Azam, Muhammad Farzik Ijaz, Rapaka S. Chandra Bose, Malini K.A, Rasmi T, Varun T.S, Yahui Xue, Gulmurza Abdurakhmanov, Dibya Prakash Rai, Gulbahor Vokhidova



First published in London, United Kingdom, 2023 by IntechOpen
IntechOpen is the global imprint of INTECHOPEN LIMITED, registered in England and Wales, registration number: 11086078, 5 Princes Gate Court, London, SW7 2QJ, United Kingdom

British Library Cataloguing-in-Publication Data
A catalogue record for this book is available from the British Library

Additional hard and PDF copies can be obtained from orders@intechopen.com

New Materials and Devices for Thermoelectric Power Generation
Edited by Basel I. Abed Ismail
p. cm.
Print ISBN 978-1-83769-508-9
Online ISBN 978-1-83769-507-2
eBook (PDF) ISBN 978-1-83769-509-6

# Meet the editor

Dr. Basel I. Abed Ismail is a successful and experienced academician and associate professor and chair of the Department of Mechanical and Mechatronics Engineering, Lakehead University, Ontario, Canada. He is a pioneering international academic researcher in alternative energy and renewable and green energy engineering technologies. He is also an academic editor for several books and an author of many articles and book chapters. Dr. Ismail is a licensed professional engineer (PEng) in Ontario, Canada.

# Contents

# Preface

Due to the increasing demands for and attraction towards alternative and green power generation, thermoelectric (TE) technology is considered to be a promising option for its inherent merits and advantages over other alternative technologies. Thermoelectrics is the direct conversion of thermal energy (heat) due to temperature differences into electrical energy using the solid-state Seebeck effect in semiconductors to attempt to mitigate the worldwide energy crisis and reduce air pollution and greenhouse gas (GHG) emissions. The key distinct advantages of using thermoelectric power generators (TEGs) include their compactness and safety, their flexibility as power sources and convenience for remote applications, and their eco-friendliness. They can also operate at high temperatures, are very reliable, and have no mechanical moving parts and therefore are noise-free in operation with few maintenance requirements. The major shortcoming of TEGs are their relatively low conversion and thermal efficiency. This has been a major cause for restricting their use in electrical power generation to certain specialized fields. However, ongoing substantial improvements in TE materials engineering, system optimization, and novel manufacturing technologies with recent advances in nanotechnology and machine learning/artificial intelligence (AI) have renewed the significance of TEGs. The design and fabrication of novel TE materials are challenging because they require co-optimization of complex properties to efficiently convert thermal energy to electricity.

This book, *New Materials and Devices for Thermoelectric Power Generation*, includes contributions from several leading researchers and experts worldwide. The topics addressed in this book are wide-ranging, from fundamentals and modeling of TE devices, new TE materials, and nanofluids for TE applications to recent strategies for improving TE devices and systems and using machine learning and computational chemistry to discover new TE materials. The book is a useful resource and adds to the existing knowledge of TE power generation applications.

I would like to sincerely thank all chapter authors for their efforts and the quality of their contributions. I would also like to thank Publishing Process Manager Ms. Zrinka Tomicic at IntechOpen for her excellent efforts in managing the publication process of this book.

**Dr. Basel I. Ismail, P.Eng.**
Associate Professors and Chair,
Department of Mechanical and Mechatronics Engineering,
Lakehead University
Thunder Bay, Ontario, Canada

Chapter 1

# Fundamental Aspects and Advances in Thermoelectric Materials for Power Generation: A Numerical Simulation Case Study

*Basel I. Abed Ismail and Jehad H. Ismail Abed*


## Abstract

Power generation using thermoelectric generator technology is becoming increasingly attractive solution due to the ongoing substantial improvements in material engineering, system optimization, and novel manufacturing technologies with recent advances in nanotechnology. The design and fabrication of novel thermoelectric materials is challenging because they require co-optimization of complex properties to efficiently convert thermal energy to electricity in what is known as the *Seebeck effect*. Computational chemistry and machine learning offer a solution toward finding optimal thermoelectric semiconductor alloys with higher figure of merit values. In this chapter, fundamental aspects and advances in thermoelectric materials for power generation are presented and discussed. A thorough modeling and numerical simulation for a case study of a TEG device application are also presented and discussed in this chapter.


**Keywords:** thermoelectric power generation, novel thermoelectric material, figure of merit, nanomaterial, clean power technology, computational chemistry, machine learning, artificial intelligence, Seebeck effect

## 1. Introduction

Due to the increasing demands and attractions toward the use of alternative and green power generation, thermoelectric (TE) technology is considered to be promising one for its inherent merits and advantages over other alternative technologies. Thermoelectrics is the direct conversion of thermal energy (heat) due to temperature difference into electrical energy using solid-state Seebeck effect in semiconductors that would contribute to mitigating the worldwide energy crisis, and reduce air pollution and GHG emissions. The key distinct merits of using thermoelectric power generator (TEGs): they are compact and safe devices; they are flexible power sources and convenient for remote applications; they are eco-friendly; they are capable of operating at high temperatures; they are very reliable; they have no mechanical moving parts and therefore noise-free in operation with significantly less maintenance



IntechOpen

requirements [1]. The major shortcoming of the thermoelectric power generator (TEG) is its relatively low conversion and thermal efficiency. This has been a major cause in restricting their use in electrical power generation to certain specialized fields. However, the ongoing substantial improvements in TE material engineering, system optimization, and novel manufacturing technologies with recent advances in nanotechnology and machine learning/artificial intelligence (AI) bring TEG to a different level of renewed significance. Over the past couple of decades, TEG applications included industrial instruments, medical and aerospace, military, and applications for mobile and remote power supply [2–4]. More recently, industrial applications involving TEG technology are becoming more attractive, especially if the waste heat associated with these applications is dissipated to the environment and available at no cost, which could be used as heat source for operating TEG and producing power at larger amounts possible [2]. In these applications, large quantities of waste heat energy are discharged into the earth's environment much of it at temperatures which are typically too low to recover using conventional electrical power generators. In general, the cost of a TEG mainly consists of the device cost and operating cost. Ismail and Ahmed [1] and Ismail [5] presented various interesting waste-heat industrial related applications where TEGs were successfully used. Luo et al. [6] presented recent advances in modeling and simulation of thermoelectric power generation. They performed a comprehensive review of theoretical models with a specific focus on the different modeling approaches and different application scenarios. In particular, the basic principles of theoretical models of the TEG were presented in their paper, including the thermal resistance model, thermal-electric numerical model, and analogy model. They also reviewed in detail the theoretical models of the TEG system, including the thermal resistance-based analytical model, computational fluid dynamics models, and fluid-thermal-electric multiphysics field coupled numerical model. In their work, the methods to improve the accuracy of theoretical models were also discussed.

In this chapter, some fundamental and material related aspects of TEG are introduced and discussed. A thorough modeling and numerical simulation for a case study of a TEG device application are also presented and discussed in this chapter.

## 2. Evolution of thermoelectric materials: from nanostructures to wearable applications

Over the past couple of decades, there has been extensive research carried out related to advances in thermoelectric materials for various applications. In general, effective TE materials should have a low thermal conductivity but a high electrical conductivity. Most widely used TE semiconductor materials are based on Bismuth Telluride ($Bi_2Te_3$), Lead Telluride (PbTe), and Si-Ge alloys. The large amount of research in thermoelectric materials has focused on optimizing the nanostructure of the thermoelectric materials to specifically improve the thermoelectric properties (e.g., increasing the Seebeck coefficient and reducing the thermal conductivity) of the TEG device. For example, Weiling and Shantung [7] reported that because the electrical conductivity and thermal conductivity correlate with the charge carriers, new methods of material synthesis must be applied to conciliate the contradiction between high electrical conductivity and low thermal conductivity. Research has been focused mainly on improving the material's figure-of-merit $Z\overline{T}$, and thus the TE conversion and thermal efficiencies, by reducing the lattice thermal conductivity [8]. Riffat and

Ma [2] and Weiling and Shantung [7] reported that thermoelectric materials, such as Lead Telluride PbTe and Bismuth Telluride $Bi_2Te_3$, have a $Z\overline{T}$ value of approximately 1.0 (at 500–700 K for PbTe and at room temperature for $Bi_2Te_3$). However, at a higher $Z\overline{T}$ values (somewhere between 2.0 and 3.0), TEG would become competitive with other power generation systems (e.g., solar power, nuclear power, wind power, fuel cell power generation systems). Very recently, Cui et al. [9] reported that $AgSbSe_2$ (Antimony-Silver-Selenium-based alloy) has a structure similar to PbTe (Lead-Tellurium-based alloy) but does not contain toxic element Pb or expensive element Te. In addition, this new alternative TE alloy has both high Seebeck coefficient and inherently low thermal conductivity, which makes it a promising candidate for medium-temperature TE power generation. Most recently, Zulkepli et al. [10] provided an overview of the key challenges in optimizing $Z\overline{T}$ values according to their TE physical properties including the state-of-the-art of the advanced approaches in $Z\overline{T}$ optimization more particularly for TE materials at low-temperature operating applications.

Moreover, it is desirable to fabricate TE modules which can conform easily to a heat source surface which would improve the thermal contact to heat sources of arbitrary geometry. Therefore, recent research has also been focused on developing novel flexible- and cylindrical-based shapes of TEG devices. For example, Yadav et al. [11] proposed and demonstrated the use of flexible and cost-effective TEG based on thin film thermoelectric on flexible fiber substrates. Min and Rowe [12] have also recently developed a novel tube-shape thermoelectric module for power generation. Most recently, Soleimani et al. [13] reported that wearable TEG devices are becoming attractive power supply for relatively low-power electronic devices. They indicated from their research that to maximize the higher power output from these devices, the focus should be on improving TE material, such as using promising electrically conductive and flexible hybrid organic and inorganic TE material, and the configuration and arrangement of thermoelements in the TEG device. In these applications, the thermal energy sources of the human body are used to power these wearable devices. Soleimani et al. [13] provided extensive research of recent studies on wearable TEG devices. Most recently, Zhu et al. [14] indicated the challenges and outlooks toward future development of wearable TEG devices and their potential applications. Lemine et al. [15] reported a comprehensive review research highlighting the promising and future-generation TEG devices based on thin film technologies with highly flexible, transparent, non-toxic, plentiful, and light-density $p$-type Copper Iodide (CuI) thin films. Sanin-Villa [16] reported that research involving carbon nanotubes/polyaniline composite films has found promising results due to its low thermal conductivity.

## 3. Accelerating the discovery of TE materials using computational chemistry and machine learning

The design of novel thermoelectric materials is challenging because they require co-optimization of complex properties to efficiently convert thermal energy to electricity in what is known as the *Seebeck effect*. As stated previously, a TE material should possess high electric conductivity to allow sufficient movement of electrons while simultaneously having low thermal conductivity to prevent heat transfer or loss. This fine balance between thermal and electrical conductivity is captured in the dimensionless $Z\overline{T}$, which is calculated using the Seeback coefficient, internal electric resistivity, and thermal conductivity of the TE material. Optimizing $Z\overline{T}$ typically requires tuning the electronic structure and various properties of the material making the design process complicated. Additionally, the quest to identify TE materials with desired properties is amplified by the sheer size of the chemical space. For a long time, discovery of TE materials has been led by *Edisonian* methods of experimental error-and-trial that have been employed to navigate the chemical space making material exploration slow and expensive. However, lately, alternative paradigms such large-scale computations have been widely adopted to accelerate the process and move beyond serendipitous discovery. Many of the research studies in this area have focused on high-throughput density functional theory (DFT) calculations to compute desired properties from a myriad selection of input atomic structures. The materials are then ranked based on their computed target property such as $Z\overline{T}$ or the power factor and the most promising candidates are sent for experimental testing. While this approach has shown significant speed ups in guiding material design, the heavy computational cost associated with DFT still limits discovery. Due to how large and diverse the search space for optimal material selection is, recently, this issue has been tackled by utilizing machine learning models. The models could learn from large DFT-curated or experimental datasets enabling access to uncharted chemical spaces while requiring only a fraction of the computational cost. Gorai et al. [17] reported that the DFT method offers an efficient and feasible route of calculating material properties replacing explicit atomic quantum calculations without significantly compromising accuracy. It works by estimating the electronic density $\rho(r)$ of the multiple electrons in the atomic structure instead of calculating the individual wave functions of electrons. This becomes increasingly important as the number of electrons and the complexity of the electron-electron interactions increase in large systems. DFT captures the effect of electron-electron interactions on the energy of the system using exchange-correlation functionals. The energy of the system could be obtained by solving Kohn-Sham's equation shown below:

$$E(\rho(r)) = \sum_j E_j + E_{xc}(\rho(r)) - \int dr\, v_{xc}(r)\rho(r) - \frac{1}{2}\int drdr' \frac{\rho(r)\rho(r)'}{|r - r'|} \tag{1}$$

where $E_j$ is the energy of Kohn-Sham orbitals, $E_{xc}$ is the exchange-correlation functional, and $v_{xc}$ is the exchange-correlation potential = $\frac{\delta E_{xc}(\rho(r))}{\delta\rho(r)}$. The equation is solved by starting with an initial guess for the charge density and then iteratively calculating the energy and updating the value of the charge density through a self-consistent scheme until convergence is achieved. The component that is most crucial to accuracy in this equation is exchange-correlation functionals. These functionals factor in for the exchange and correlation energies of the electron interactions. While their energy values make up a small portion of the total energy of the system, careful choice of functionals is a key to predict the electronic structure and properties of different classes of materials with sufficient accuracy for experimental verification.

Local density approximation (LDA) is one of the simplest functionals that were implemented for DFT calculations. It neglects spatial variations of the electron density across the system and only considers local correlation effects by treating charge density as a uniform gas of electrons, which offers a reasonable accuracy for some applications [18]. To incorporate non-local correlations, generalized gradient approximation (GGA) improves on LDA by considering not only the electron density at a given point but also considers the gradient of the electron density across the space. This provides a more accurate model accounting for electron density variations that are important in many systems. There has been a growing interest in developing more accurate functionals such as hybrid functionals and machine learning (ML)-based functionals capture complex correlations beyond the scope of LDA and GGA [19]. One of the most popular functionals used for screening of thermoelectric materials using DFT is PBE-GGA [20] and van der Waals-corrected functionals DFT-D2 [21]. However, despite selecting a suitable functionals, there are still inherit challenges that stem from the nature of DFT calculations, which do not account for temperature effects. Additionally, the lack of standardization of functionals prevents the direct comparison and transfer of results in the literature. This in turn widens the gap between computational predictions and experimental validations. Choubisa et al. [22] addressed both issues listed above regarding computational cost and accuracy of predictions by utilizing an error-correction learning (ECL) based on a neural network to model the error correction function from experimental data. This approach is implemented in two basic steps: (1) The model learns from prior experimental or computational datasets and then (2) it utilizes new experiments to provide feedback and refine the model to improve its accuracy. The second step allows the model to implicitly account for disparities in synthesis methods, material morphology, and defects, which vary from one lab setting to another and are normally not captured by DFT calculations. This improved model is then used to screen the material space, particularly Materials Project [23], a large open computational dataset of materials, with the purpose of discovering new promising chemistries. The authors focused their search on low-temperature thermoelectric materials <300°C. Notably, a new chemical family based on PbSe:SnSb (lead selenide:tin antimony) of thermoelectric materials was identified using this approach. The best composition exhibited double the power factor of a standard PbSe (lead selenide). This study encourages the development of new hybrid strategies guided by computational and experimental results, as well as machine learning models to shorten the cycle of developing thermoelectric materials.

## 4. The principle of a thermoelectric generator (TEG)

**Figure 1** shows how a thermoelectric generator (TEG) works. A temperature difference is established between two junctions, namely, the hot junction and the cold junction, of two dissimilar materials made of metals or semiconductors. Due to this temperature difference, a voltage is generated using the Seebeck effect. This effect is fundamentally used in applications of thermocouples for temperature measurements. TEG devices can operate as electrical power generators using this effect. In a basic TEG thermocouple, heat transfers at a rate $\dot{Q}_H$ from the heat source maintained at a high temperature $T_H$ to the hot junction. It then transfers through the thermoelectric materials A and B and reaches the cold junction maintained at a low temperature $T_H$ where heat is rejected at the heat sink at a rate $\dot{Q}_L$. The heat transferring at the hot junction causes an electric current to flow in the circuit based on Seebeck effect and an electrical power is produced when an electric load is connected to this circuit. This constitutes a thermodynamic heat engine with a power cycle established. The electrons (charge carriers) serve as the working fluid. The power output $\dot{W}_e$ is the difference between $\dot{Q}_H$ and $\dot{Q}_L$based on the first law of thermodynamics.

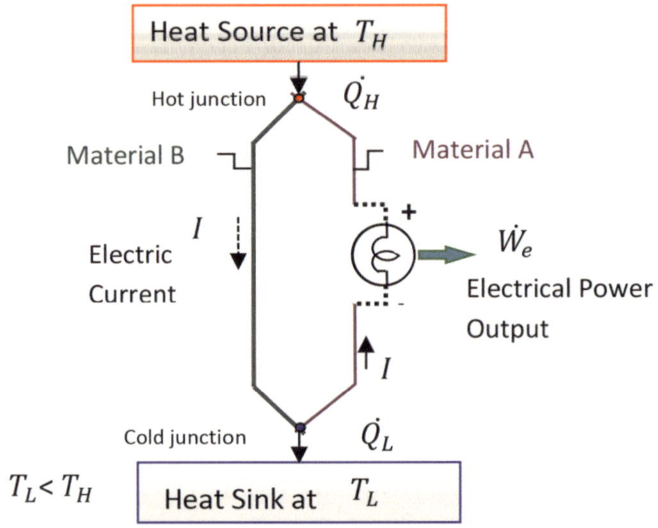


**Figure 1.**
*A schematic diagram showing the principle of a TEG.*

**Figure 2** shows a schematic diagram of a simple TEG device with its arrangement of components [1]. The TEG device is composed of two ceramic plates (substrates) that serve as a foundation, providing mechanical integrity, and electrical insulation for $n$-type (heavily doped to create excess electrons) and $p$-type (heavily doped to create excess holes) semiconductor thermoelements. In TE materials, electrons and holes operate as both charge carriers and energy carriers. There are very few modules designed and fabricated without ceramic plates. Removing the ceramic plates from the

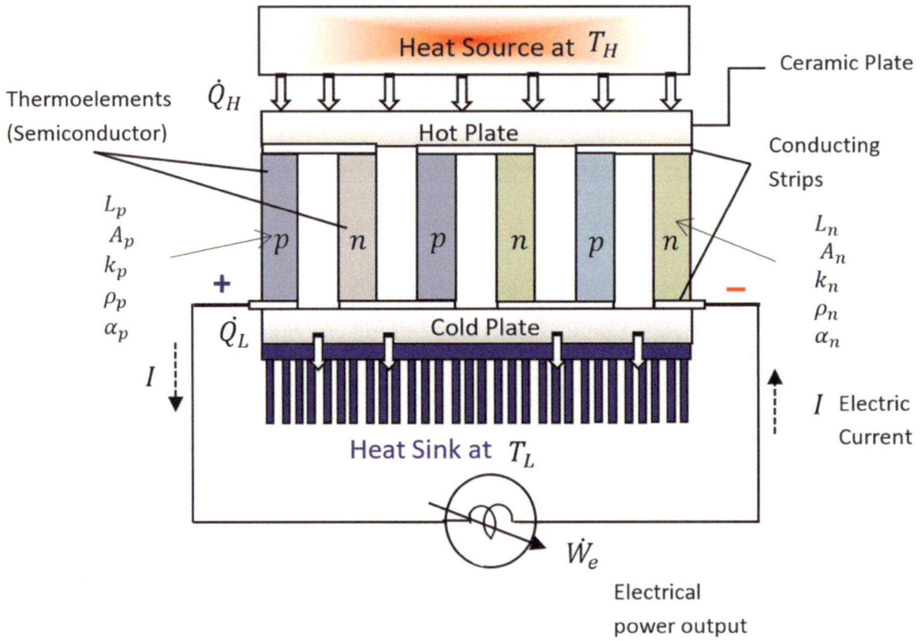


**Figure 2.**
*A schematic diagram showing arrangement of a basic TEG device.*

modules would eliminate the thermal resistance associated with them but might lead to mechanical fragility of the module. The ceramic plates are commonly made from alumina, but when large lateral heat transfer is required, materials with higher thermal conductivity (e.g., beryllia and aluminum nitride) are desired. The semiconductor thermoelements (e.g., silicon-germanium, lead-telluride-based alloys) that are sandwiched between the ceramic plates are connected thermally in parallel and electrically in series to form a thermoelectric device (module). More than one pair of semiconductors are normally assembled together to form a thermoelectric module and within the module a pair of thermoelements is called a thermocouple.

The junctions connecting the thermoelements between the hot and cold plates are interconnected using highly conducting metal (e.g., copper) strips as shown in **Figure 2**. The power output for most of the commercially available TEGs ranges from microwatts to multi-kilowatts [2] (Rowe, 1999).

## 5. Modelling and numerical simulation: case study of the performance of a TEG device

In **Figure 1**, the heat transfer at the hot junction, $\dot{Q}_H$ includes three terms, given by [24]:

$$\dot{Q}_H = \dot{Q}_{SE} + \dot{Q}_{JH} + \dot{Q}_{CON} \tag{2}$$

where $\dot{Q}_{SE}$ is heat flow due to Seebeck effect, given by

$$\dot{Q}_{SE} = \alpha \, T_H \, I \tag{3}$$

$\dot{Q}_{JH}$ is the irreversible heat flow due to Joule heating effect that is generated as the electric current flows in the wire. There are two elements (legs) of the TEG module, so this heat on each leg accounts for one-half of Jule heating, given by [24]:

$$\dot{Q}_{JH} = -\frac{1}{2} I^2 R \tag{4}$$

$\dot{Q}_{CON}$ is heat flow associated with conduction heat transfer, given by

$$\dot{Q}_{CON} = K \, (T_H - T_L) \tag{5}$$

Substituting Eqs. (3)–(5) in Eq. (2), yields

$$\dot{Q}_H = \alpha \, T_H \, I - \frac{1}{2} I^2 R + K \, (T_H - T_L) \tag{6}$$

In Eq. (6), α is the Seebeck coefficient, $R$ is the internal electrical resistance, and $K$ is the total thermal conductance of the thermoelements, given by

$$R = \frac{\rho \, L}{A} \tag{7}$$

$$R = R_p + R_n \tag{8}$$

$$K = \frac{k\,A}{L} \tag{9}$$

$$K = K_p + K_n \tag{10}$$

where $A$, $L$, $k$ and $\rho$ are the cross-sectional area, length, thermal conductivity, and electrical resistivity' of the $n$-type and $p$-type thermoelements, respectively. The Seebeck coefficient $\alpha$ for the TE device can be rewritten in terms of the average values of Seebeck coefficients for the dissimilar $n$-type and $p$-type thermoelements as [24]

$$\alpha = \alpha_p - \alpha_n \tag{11}$$

The performance of thermoelectric materials can be expressed by figure of merit, $Z$, given by

$$Z = \alpha^2/KR \tag{12}$$

The relationship of $K$ and $R$ with properties of the thermoelements is given by

$$K\,R = \left[\left(k_p\rho_p\right)^{1/2} + \left(k_n\rho_n\right)^{1/2}\right]^2 \tag{13}$$

Maximizing $Z$ means that $K\,R$ should be minimized. The minimum $K\,R$ necessitates that

$$\frac{L_n}{A_n} = \frac{L_p}{A_p}\sqrt{\frac{k_n\rho_p}{k_p\rho_n}} \tag{14}$$

The power generated at the electrical load $\dot{W}_e$ can be determined using,

$$\dot{W}_e = I^2\,R_L \tag{15}$$

where $R_L$ is the load electrical resistance. The thermal efficiency of a TEG device (treated as a thermodynamic heat engine), $\eta_{th}$, is defined as

$$\eta_{th} = \frac{\dot{W}_e}{\dot{Q}_H} \tag{16}$$

Eq. (16) can be rewritten using Eqs. (6) and (15) as

$$\eta_{th} = \frac{I^2\,R_L}{\alpha\,T_H\,I + \frac{1}{2}I^2R + K\,(T_H - T_L)} \tag{17}$$

Using the first law of thermodynamics of the closed system, the heat transfer at the low-temperature side (cold junction), $\dot{Q}_L$ can be written as

$$\dot{Q}_L = \dot{Q}_H - \dot{W}_e \tag{18}$$

In **Figure 1**, the potential difference (or voltage $V$) is proportional to the temperature difference $\Delta T$, given by

$$V = \alpha \Delta T \tag{19}$$

where,

$$\Delta T = T_H - T_L \tag{20}$$

Eq. (19) can also be written in terms of the open circuit voltage, $V_{oc}$,

$$V_{oc} = \alpha \Delta T \tag{21}$$

Also, $V_{oc}$ is given by

$$V_{oc} = I\,(R + R_L) \tag{22}$$

Equating Eqs. (21) and (22) and solving for the electric current, yield

$$I = \frac{\alpha \Delta T}{(R + R_L)} \tag{23}$$

For a thermodynamic heat engine, the maximum thermal efficiency is limited by the second law of thermodynamics given by Carnot efficiency:

$$\eta_{CAR} = 1 - \frac{T_L}{T_H} \tag{24}$$

Introducing the dimensionless parameters, θ, ξ, and β given by

$$\theta = R_L/R \tag{25}$$

$$\xi = Z\overline{T} \tag{26}$$

$$\beta = T_L/T_H \tag{27}$$

$$\eta_{CAR} = 1 - \beta \tag{28}$$

where the average temperature, $\overline{T}$, of the TE device is given by

$$\overline{T} = \frac{T_H + T_L}{2} \tag{29}$$

Substituting Eq. (23) into Eq. (17) with Eqs. (12), (24)-(26), yields

$$\eta_{th} = \eta_{CAR}\left[\frac{\theta}{\frac{(1+\theta)^2}{Z\,T_H} + (1+\theta) - \left(\frac{\eta_{CAR}}{2}\right)}\right] \tag{30}$$

The product $Z\,T_H$ in Eq. (30) can be written in terms of the dimensionless parameters ξ, and β, as

$$Z\,T_H = \frac{2\,\xi}{(1+\beta)} \tag{31}$$

Substituting Eqs. (27) and (31) into Eq. (30), gives the thermal efficiency of the TE device in terms of dimensionless parameters, θ, ξ, and β, as

$$\eta_{th} = \frac{\theta\, \eta_{CAR}}{(1+\theta) - \left(\frac{\eta_{CAR}}{2}\right) + \frac{1}{2\xi}(1+\theta)^2(1+\beta)} \tag{32}$$

Normally, in designing and operating TEGs for various power generation applications, it would be very useful to maximize their performance. To determine the maximum power generation and conversion efficiency of the TEG module, the following set of equations is developed.

The maximum efficiency of the TE device can be determined using

$$\frac{d\eta_{th}}{d\theta} = 0 \tag{33}$$

Which leads to

$$\theta_{max} = \sqrt{(1+\xi)} \tag{34}$$

Using Eq. (34) in Eq. (32), yields

$$\eta_{th,\,max} = \eta_{CAR}\left[\frac{\theta_{max} - 1}{\theta_{max} + \beta}\right] \tag{35}$$

Substituting Eq. (23) in Eq. (15) and using Eq. (26), gives

$$\dot{W}_e = \theta\left(\frac{\alpha \Delta T}{\theta + 1}\right)^2 \tag{36}$$

The maximum power output of the TE device can be determined using

$$\frac{d\dot{W}_e}{d\theta} = 0 \tag{37}$$

This leads to the condition of maximum power of the TEG device,

$$\theta = 1 \tag{38}$$

That is, $R_L = R$.

In this case, we have

$$\dot{W}_{e,\,max} = I_{mp}^2 R \tag{39}$$

$$I_{mp} = \frac{\alpha \Delta T}{2R} \tag{40}$$

Eq. (39) can then be rewritten as

$$\dot{W}_{e,\,max} = \frac{1}{R}\left(\frac{\alpha \Delta T}{2}\right)^2 \tag{41}$$

Also, the maximum power output can be expressed in terms of

$$\dot{W}_{e,\,max} = I_{mp}\, V_{mp} \tag{42}$$

where,

$$V_{mp} = \frac{\alpha \Delta T}{2} \tag{43}$$

The maximum current can be determined by setting $R_L = 0$ in Eq. (23), as

$$I_{max} = \frac{\alpha \Delta T}{\mathrm{R}} \tag{44}$$

The maximum voltage is the open circuit voltage, using Eq (21) (or Eq. (22))

$$V_{max} = \alpha \Delta T \tag{45}$$

The maximum power efficiency is given by

$$\eta_{mp} = \eta_{CAR} \left[ \frac{1}{\frac{4}{ZT_H} - \frac{\eta_{CAR}}{2} + 2} \right] \tag{46}$$

For a TE device with λ multicouple $n$-type and $p$-type thermoelements in a TEG device, we have

$$\left(\dot{Q}_H\right)_\lambda = \lambda \dot{Q}_H \tag{47}$$

$$\left(\dot{W}_e\right)_\lambda = \lambda\, \dot{W}_e \tag{48}$$

$$(V)_\lambda = \lambda\, V \tag{49}$$

$$(K)_\lambda = \lambda\, K \tag{50}$$

$$(R)_\lambda = \lambda\, R \tag{51}$$

$$(R_L)_\lambda = \lambda\, R_L \tag{52}$$

$$(I)_\lambda = I \tag{53}$$

$$(\eta_{th})_\lambda = \eta_{th} \tag{54}$$

Numerical simulations: *case study*

A TEG module is to be designed with its performance analytically and numerically evaluated to deliver a total electrical power of 1 kW using a waste heat source from hot exhaust gas at 811 K produced from an IC engine. The cold heat is rejected at a temperature of 436 K. The design specifications, material properties, and thermal conditions of the $p$-type and $n$-type of the PbTe thermoelements of the TEG device are summarized in **Table 1**. For this practical case study, it is required to:

A. Determine the maximum thermal efficiency, the efficiency at maximum power output, the number of thermocouples required for the TEG device, and other performance parameters.

B. Perform numerical simulations to determine the effect of lowering the cold temperature (heat sink temperature) on the key performance parameters of the TEG device.

| | |
|---|---|
| **Design specifications** | |
| $L_p$ | 8 mm |
| $L_n$ | 8 mm |
| $A_n$ | 0.6 $cm^2$ |
| **TE material properties: Lead Telluride (PbTe)** | |
| $k_n$ | 0.0140 W/cm K |
| $k_p$ | 0.0120 W/cm K |
| $\rho_n$ | 0.00101 Ω cm |
| $\rho_p$ | 0.00095 Ω cm |
| $\alpha_n$ | $-170 \times 10^{-6}$ V/K |
| $\alpha_p$ | $190 \times 10^{-6}$ V/K |
| **Thermal conditions** | |
| $T_L$ | 436 K |
| $T_H$ | 811 K |

**Table 1.**
*Design specifications, material properties, and thermal conditions for the TEG device.*

TEG device performance parameters were calculated using MS Excel program. The numerical results are shown in **Table 2**. For this case, 593 thermocouples are required to construct the TEG device to deliver 1 kW of electrical power. The maximum thermal efficiency and the efficiency at maximum power output of the TEG device were found to be 13.1% and 12.5%, respectively.

The numerical simulations were carried out using the simple computer program and the numerical results are shown in **Table 3**. It is interesting to see the detrimental effect of lowering the heat sink temperature of the TEG device on the various efficiency values of the device. Also, the number of multi-thermocouple required in constructing the TEG device decreased significantly with decreasing the heat sink temperature of the device. For example, by operating the device at the cold temperature of room temperature of 25°C, there are 317 thermocouples (i.e., $\lambda = 317$) required in constructing the device as opposed to 593 thermocouples when operating the TEG device at 163°C, as can be determined using Eq. (48). The difference of 276 reflects a significant drop in the required material for designing and building this device which would then significantly reduce the cost of the device and its maintenance cost requirements. The efficiency at the maximum power output of the TEG device increased from 12.5% to 17.5% (i.e., by 5% increase) only by means of lowering the heat sink temperature to room temperature. It should be noted that lowering the cold heat sink operating temperature $T_L$ would increase the potential temperature difference (driving force for TEG device) $\Delta T$ at a fixed heat source temperature $T_H$, as can be seen in Eq. (20). This in turn would result in increasing the maximum power output of the TEG device (see Eq. (41)) per thermocouple. Ultimately, this would result in lowering the number of thermocouples (i.e., thermoelements $\lambda$) required to form the TEG device, as can be seen in Eq. (48), for a given total power output requirement in an application where TEG device is used. One another useful operation

| Parameter | Value | Equation # used |
|---|---|---|
| $\alpha$ | $360 \times 10^{-6}$ V/K | Eq. (11) |
| $\beta$ | 0.538 | Eq. (27) |
| $e$ | 0.629 $cm^2$ | Eq. (14) |
| $R$ | 0.002555 Ω | Eqs. (7) and (8) |
| $K\,R$ | $5.0927 \times 10^{-5}$ ΩW/K | Eq. (13) |
| $Z$ | 0.002545 $K^{-1}$ | Eq. (12) |
| $\overline{T}$ | 623 K | Eq. (29) |
| $\xi$ | 1.586 | Eq. (26) |
| $\theta_{max}$ | 1.6081 | Eq. (34) |
| $R_L$ | 0.004109 Ω | Eq. (25) |
| $\theta$ | 1.6082 | Eq. (25) |
| $\eta_{CAR}$ | 0.462 | Eq. (28) |
| $\eta_{th,\,max}$ | 0.1309 | Eq. (35) |
| $Z\,T_H$ | 2.0624 | Eq. (31) |
| $I$ | 20.26 A | Eq. (23) |
| $\dot{W}_e$ | 1.6866 W | Eq. (15) |
| $\eta_{th}$ | 0.1309 | Eq. (30) |
| $\dot{Q}_H$ | 12.8846 W | Eq. (16) |
| $\dot{Q}_L$ | 11.1980 W | Eq. (18) |
| $\alpha\Delta T$ | 0.135 V | |
| $I_{max}$ | 52.84 A | Eq. (44) |
| $V_{max}$ | 0.135 V | Eq. (45) |
| $I_{mp}$ | 26.42 A | Eq. (40) |
| $V_{mp}$ | 0.0675 V | Eq. (43) |
| $\dot{W}_{e,\,max}$ | 1.7833 W | Eq. (42) |
| $\eta_{mp}$ | 0.1246 | Eq. (46) |
| $(\dot{W}_e)_\lambda$ | 1000 W | Given |
| $\lambda$ | 593 thermocouples | Eq. (48) |
| $(\dot{Q}_H)_\lambda$ | 7641 W | Eq. (47) |
| $(\dot{Q}_L)_\lambda$ | 6641 W | Eq. (18) |

**Table 2.**
*Calculations of the TEG performance parameters.*

metric is the size of the TEG device. As the number of thermocouples is decreased, the volume and weight of the device would decrease. This will make the TEG device more compact and cost-effective.

| $T_L$ (°C) | 163 | 150 | 125 | 100 | 75 | 50 | 25 (room temp) |
|---|---|---|---|---|---|---|---|
| $\eta_{CAR}$ (%) | 46.2 | 47.8 | 50.9 | 54.0 | 57.1 | 60.2 | 63.3 |
| $\eta_{th,max}$ (%) | 13.1 | 13.6 | 14.5 | 15.4 | 16.3 | 17.3 | 18.2 |
| $\eta_{mp}$ (%) | 12.5 | 12.9 | 13.8 | 14.7 | 15.6 | 16.5 | 17.5 |
| λ | 593 | 554 | 489 | 435 | 389 | 350 | 317 |

**Table 3.**
*Numerical simulation results (case study—heat source exhaust gas temperature $T_H$ = 538°C).*

## 6. Conclusion

In this chapter, some fundamental and material related aspects of thermoelectric materials are introduced and discussed. A thorough modeling and numerical simulation for a hypothetical case of a TEG device are also presented and discussed in this chapter. The numerical simulation was carried out for a hypothetical practical case of a TEG made of PbTe semiconductors alloy. It was found that the heat sink cold temperature has a detrimental effect on the thermal and conversion efficiencies of the TEG device. More particularly, the efficiency at the maximum power output of the TEG device increased from 12.5% to 17.5% (i.e., by 5% increase) only by means of lowering the heat sink temperature to room temperature. In addition, the cost of the TEG device and its maintenance cost requirements were significantly reduced by lowering the operating cold temperature of the device for this case study.

## Acronyms

| | |
|---|---|
| DFT | density function theory |
| GHG | greenhouse gases |
| GGA | generalized gradient approximation |
| LDA | local density approximation |
| ML | machine learning |
| TE | thermoelectric |
| TEG | thermoelectric generator |

## Author details

Basel I. Abed Ismail[1]* and Jehad H. Ismail Abed[2]

1 Department of Mechanical and Mechatronics Engineering, Lakehead University, Thudner Bay, Ontario, Canada

2 Artificial Intelligence, Meta, Toronto, Ontario, Canada

*Address all correspondence to: baseliai@gmail.com

## References

[1] Ismail B, Ahmed WH. Thermoelectric power generation using waste-heat energy as an alternative green technology. Recent Patents on Electrical Engineering. 2009;**2**(1):27-39

[2] Riffat SB, Ma X. Thermoelectrics: A review of present and potential applications. Applied Thermal Engineering. 2003;**23**:913-935. ISSN: 1359-4311

[3] Rowe DM, Min G. Evaluation of thermoelectric modules for power generation. Journal of Power Sources. 1998;**73**:193-198. ISSN: 0378-7753

[4] Stevens JW. Optimal design of small $\Delta T$ thermoelectric generation systems. Energy Conversion and Management. 2001;**42**:709-720. ISSN: 0196-8904

[5] Ismail BI. Automotive exhaust gas waste-heat recovery for green electrical power generation using thermoelectric technology. Recent Patents on Electrical Engineering. 2012;**5**:185-197

[6] Luo D, Liu Z, Li Y, Wang R, Zhang L, Yang X. Recent advances in modeling and simulation of thermoelectric power generation. Energy Conversion and Management. 2022;**273**:1-29

[7] Weiling L, Shantung TU. Recent developments of thermoelectric power generation. Chinese Science Bulletin. 2004;**49**(12):1212-1219. ISSN: 1001-6538

[8] Rowe DM. Thermoelectric waste heat recovery as a renewable energy source. International Journal of Innovations in Energy Systems and Power. 2006;**1**:13-23

[9] Cui J, Tang X, Tan G. Advances and challenges of $AGSbSe_2$-based thermoelectric materials. ChemNanoMat. 2022;**8**:1-8

[10] Zulkepli N, Yuinas J, Mohamed MA, Hamzah AA. Review of thermoelectric generators at low operating temperatures: Working principles and materials. Micromachines. 2021;**12**:2-25

[11] Yadav A, Pipe KP, Shtein M. Fiber-based flexible thermoelectric power generator. Journal of Power Sources. 2008;**175**:909-913. ISSN: 0378-7753

[12] Min G, Rowe DM. Ring-structured thermoelectric module. Semiconductor Science and Technology. 2007;**22**: 880-883. ISSN: 0268-1242

[13] Soleimani Z, Zoras S, Ceranic B, Cui Y, Shahzad S. A comprehensive review on the output voltage/power of wearable thermoelectric generators concerning their geometry and thermoelectric materials. Nano Energy. 2021;**89**:2-31

[14] Zhu S, Fan Z, Feng B, Shi R, Jiang Z, Peng Y, et al. Review on wearable thermoelectric generators: From devices to applications. Energies. 2022;**15**:2-27

[15] Lemine AS, Bhadra J, Al-Thani NJ, Afmad Z. Promising transparent and flexible thermoelectric modules based on p-type Cul thin films – A review. Energy Reports. 2022;**8**:11607-11637

[16] Sanin-Villa D. Recent developments in thermoelectric generation: A review. Sustainability. 2022;**14**:2-20

[17] Gorai P, Stevanovic V, Toberer ES. Computationally guided discovery of thermoelectric materials. Nature Reviews Materials. 2017;**2**(9):1-16

[18] Becke AD. Perspective: Fifty years of density-functional theory in chemical physics. Journal of Chemical Physics. 2014;**140**(18):18A301

[19] Krukau AV, Vydrov OA, Izmaylov AF, Scuseria GE. Influence of the exchange screening parameter on the performance of screened hybrid functionals. Journal of Chemical Physics. 2006;**125**(22):224106

[20] Perdew JP, Burke K. General gradient approximation made simple. Physical Review Letters. 1996;**77**:3865

[21] Grimme S. Semiempirical GGA-type density functional constructed with a long-range dispersion correction. Journal of Computational Chemistry. 2006;**27**(1787):1787-1799

[22] Choubisa H, Haque MA, Zhu T, Zeng L, Vafie M, Baran D, et al. Closed-loop error correction learning accelerates experimental discovery of thermoelectric materials. Advanced Materials. 2023;**35**(40):2302575. DOI: https://arxiv.org/abs/2302.13380

[23] Jain A et al. Commentary: The materials project: A materials genome approach to accelerating materials innovation. APL Materials. 2013;**1**:11002

[24] Lee H-S. Thermal Design: Heat Sinks, Thermoelectrics, Heat Pipes, Compact Heat Exchangers, and Solar Cells. Inc, Hoboken, New Jersey: John Wiley and Sons; 2010

Chapter 2

# Thermoelectric Generators: Design, Operation, and Applications

*Baljit Singh Bhathal Singh*


## Abstract

This chapter offers a comprehensive analysis of thermoelectric generators (TEGs), with a particular emphasis on their many designs, construction methods, and operational processes, all aimed at achieving optimal conversion of thermal energy into electrical energy. This chapter extensively examines the fundamental principles that control thermoelectric generators (TEGs), providing a complete examination of their respective merits and drawbacks in comparison with conventional energy conversion techniques. This study thoroughly investigates the key elements that have a significant impact on the performance of thermoelectric generators (TEGs), including the temperature gradient, heat source temperature, and load resistance. Moreover, the chapter explores the diverse range of thermoelectric materials employed in these generators and their significant qualities that directly affect the efficiency and power output of the devices. TEGs have been widely examined in terms of their practical applications, which include waste heat recovery, space exploration, and remote power generation. This chapter provides a comprehensive analysis of the obstacles and prospects associated with the incorporation of thermoelectric generators (TEGs) into renewable energy systems. Additionally, it evaluates the feasibility of scaling up TEG manufacturing to meet growing energy demands, with a specific focus on promoting sustainable energy solutions.


**Keywords:** heat energy, efficiency, power, thermoelectric, Seebeck

## 1. Introduction

### 1.1 Thermoelectric generating systems: History and importance

TEGs are solid-state devices that use the thermoelectric effect to transform thermal energy into electrical power [1]. The Seebeck effect, which happens when a temperature gradient is introduced across incompatible semiconductor materials, provides the basis for this phenomenon. Waste heat recovery and portable power production are only two of the many areas where TEGs have found recent success.

TEGs are useful because they can be used to recover heat otherwise lost in a variety of manufacturing and electricity-producing operations [2]. Exhaust gases from vehicles and industrial machinery are two common examples of waste heat sources that escape into the environment and cause damage. In order to improve energy efficiency and decrease emissions of greenhouse gases, TEGs present a possible alternative by capturing and converting this waste heat into usable electrical energy.

As solid-state devices with no moving parts, TEGs provide great dependability, compact size, and adaptability to a wide range of operating situations [3]. They are perfect for out-of-the-way places because of their quiet operation, low need for upkeep, and lengthy service life.

### 1.2 Aims of this chapter

The purpose of this chapter is to furnish readers with a thorough comprehension of thermoelectric generators (TEGs) through an examination of their structure, functioning, and wide-ranging uses. The primary objectives of this chapter are to provide a comprehensive understanding of the operational principles of thermoelectric generators (TEGs), with a specific emphasis on the Seebeck effect and thermoelectric effects. Furthermore, the chapter explores the diverse thermoelectric materials employed in thermoelectric generators (TEGs) and their corresponding features. Another important feature that is explored in this study is the exploration of the parameters that influence the performance and efficiency of TEGs. These aspects include temperature gradients, material characteristics, and device design. In addition, the chapter provides an extensive discussion of various design considerations and optimisation methodologies aimed at improving the performance of thermoelectric generators (TEGs). This chapter presents a thorough examination of the various uses of thermoelectric generators (TEGs), including waste heat recovery, portable power generation, and remote sensing. In conclusion, the chapter discusses the present obstacles and possible prospects within the sector, placing particular attention on developments in materials, device engineering, and system integration. The primary objective of this chapter is to enhance the comprehension of thermoelectric generators as a viable and effective technology for energy conversion with a focus on sustainability.

## 2. Operations principles of thermoelectric generator operation

### 2.1 The generation of electricity from thermal energy

The Seebeck phenomenon, in which a temperature difference between two dissimilar materials causes a voltage potential difference, is the basis for thermoelectric generators' operation [4]. A TEG module is made up of a series or parallel connection of many thermocouples, each of which is made up of p-type and n-type semiconductors with opposite charge carriers.

A temperature gradient is created throughout the module when one end of the thermocouple is placed in contact with a heat source (hot side), and the other end is placed in contact with a heat sink (cold side). The voltage potential difference between the hot and cold ends of each thermocouple is directly proportional to the temperature gradient between them. The stored voltage can be used to power electronics or used later.

### 2.2 How it stacks up against other energy conversion techniques

There are benefits and drawbacks to using thermoelectric generators instead of more traditional energy conversion technologies.

- TEGs' unique features excel in applications where low-power generation or waste heat recovery is critical despite their lower conversion efficiency than conventional power generation methods [5].

- Scalability: TEGs' adaptability extends to a wide range of applications, from tiny devices to power sensors to massive installations to recycle heat from factories' waste.

- Because they are solid-state devices with no moving components, emit no greenhouse gases, and run quietly, TEGs have a negligible effect on the environment.

- TEGs can function in a wide temperature range, making them adaptable for use in a variety of waste heat capture applications. However, they are extremely sensitive to the temperature gradient, with larger differentials improving performance.

### 2.3 The pros and cons of using thermoelectric power plants

Thermoelectric generator advantages include:

- Waste Heat Recovery: TEGs offer a practical approach to reusing the heat that would otherwise be wasted in several contexts, including manufacturing, power generation, and transportation [4].

- Portability and dependability: thanks to its small size, low weight, and lack of moving components, TEGs can be taken to areas that do not have access to regular power sources.

- TEGs are ideal for applications where accessibility and maintenance are difficult because of their long operational lifespan and low maintenance needs, both of which stem from their solid-state construction.

Thermoelectric generators have several drawbacks.

- The current efficiency of TEGs in converting energy is lower than that of more conventional approaches. Researchers are working to overcome this barrier by improving thermoelectric materials and optimising device design.

- The effectiveness of TEGs declines at higher temperatures, which can restrict their use in high-temperature situations, and they function best within certain temperature ranges.

- TEG performance is highly dependent on thermoelectric materials, which can be both costly and scarce. To encourage wider deployment of TEGs, researchers are looking for cost-effective and efficient materials.

## 3. Factors affecting thermoelectric generator performance

The performance of a thermoelectric generator (TEG) can be influenced by a multitude of factors. The enhancement of a thermoelectric generator's performance and power generation capabilities can be achieved through the optimisation of these aspects.

### 3.1 The influence of temperature gradient

The temperature gradient is the term used to describe the disparity in temperature between the hot and cold sides of a thermoelectric generator. There exists a direct association between the temperature gradient and the power output of a thermoelectric generator (TEG). Increased power generation is achieved through a larger temperature differential existing between the hot and cold sides of the thermoelectric generator (TEG). Heat transfer can occur from the region with a higher temperature to the region with a lower temperature, and the thermoelectric generator (TEG) has the capability to convert this heat flux into electrical energy. The Thermoelectric Generator (TEG) has the potential to generate a higher amount of electrical power as a result of an increased heat flow resulting from a larger temperature differential. The increase in temperature across the thermoelectric generators (TEGs) leads to a corresponding increase in the maximum power output [6]. The voltage output of a thermoelectric generator (TEG) is influenced by the temperature gradient. The operation of a thermoelectric generator (TEG) relies on the Seebeck effect, which generates an electric potential across the TEG in the presence of a temperature gradient. A larger temperature gradient results in an increased voltage output. The TEG has the potential to generate a higher amount of electrical power as a result of an increased heat flow resulting from a larger temperature differential [7].

### 3.2 The impact of heat source temperature on a system

The performance of a thermoelectric generator (TEG) is significantly impacted by a key parameter known as the temperature of the heat source. The temperature of the heat source significantly affects the power generation capability of a thermoelectric generator (TEG). The power generation of a thermoelectric generator (TEG) is directly influenced by the temperature gradient between its hot and cold sides. An elevated heat source temperature leads to an augmented temperature gradient, hence yielding a greater temperature disparity and an increased power output. Consequently, elevating the temperature of the heat source has the potential to enhance the power output of the thermoelectric generator (TEG). Based on the results obtained, it can be concluded that the utilisation of high input energy, coupled with a heat collector possessing a high absorptivity and low emissivity, yields advantageous outcomes in terms of generating a thermoelectric generator with superior performance [8].

### 3.3 The impact of load resistance

The load resistance plays a critical role in determining the power output and efficiency of a thermoelectric generator (TEG) system. The voltage and current produced by the thermoelectric generator (TEG) are contingent upon the load resistance. Ohm's eq. ($V = I R$) establishes a direct relationship between the voltage across the load resistance, the current flowing through it, and the value of the load resistance itself. The adjustment of voltage and current levels to align with the specific requirements of an application can be achieved through the selection of different load resistance values. The study aims to ascertain the optimal electrical load resistance that maximises thermoelectric generation in a liquid-to-liquid generator. The findings provide insights into the thermoelectric characteristics of the generator when the electrical load resistance is progressively increased [9]. The power output is inversely affected by the sum of the internal resistance and the load resistance.

*DOI: http://dx.doi.org/10.5772/intechopen.1002722*

## 4. Thermoelectric materials for generators

Thermoelectric generators (TEGs) employ thermoelectric materials to convert waste heat into electrical power. These materials possess distinct attributes that enable them to generate electricity via the Seebeck effect, wherein a voltage differential arises from a temperature gradient across the material.

### 4.1 There exist various classifications of thermoelectric materials

Bismuth telluride ($Bi_2Te_3$) is the thermoelectric material most employed for applications at room temperature. The high thermoelectric figure of merit (ZT) within the range of 1–1.5 contributes to its effectiveness in power generation. The performance of the material can be enhanced through the process of alloying with chemicals such as antimony (Sb). Commercially accessible thermoelectric generators (TEGs) that utilise $Bi_2Te_3$ semiconductors are the most cost-effective options [10]. Lead telluride (PbTe) has favourable thermoelectric properties when subjected to elevated temperatures. Due to its low thermal conductivity and significant Seebeck coefficient, this material exhibits potential for use in high-temperature applications. The performance of the material can be improved with the addition of alloying elements such as antimony (Sb) or selenium (Se) [11]. Skutterudites have exhibited promising thermoelectric properties, particularly at elevated temperature conditions. Illustrative instances encompass CoSb3 and filled skutterudites, such as (Co, Fe) $Sb_3$ with incorporated guest atoms. The high ZT values exhibited by these materials can be attributed to their intricate crystal structure and unique electrical properties. Silicides that exhibit excellent thermoelectric performance at elevated temperatures encompass Mg2Si and $Ca_3Co_4O_9$. The crystals possess unique properties pertaining to both their crystalline structure and electrical characteristics, which contribute to their high efficiency in converting energy [12]. TiNiSn and ZrNiSn are two exemplary instances of half-Heusler compounds that have exhibited considerable potential as thermoelectric materials, specifically in the context of high-temperature implementations. The thermoelectric performance of these materials is attributed to their notable electrical conductivity and moderate heat conductivity (**Table 1**).

| Operating Temperature, °C | Type | Materials | Maximum *ZT* |
|---|---|---|---|
| < 150 | P | $Bio_{0.5}Sb_{1.5}Te_3$ | 1.4 |
| | n | $Bi_2Se_{0.3}Te_{2.7}$ | 1.0 |
| | p,n | $Bi_2Te_3$ | 0.8 |
| 150–500 | P | $Zn_4Sb_3$ | — |
| | p,n | PbTe | 0.7–0.8 |
| | P | TeAgGeSb | 1.2 |
| 500–700 | P | $CeFe_4Sb_2$ | 1.1 |
| | ti | $CoSb_3$ | 0.8 |
| 700–900 | p,n | SiGe | 0.6–1.0 |
| | p | LaTe | 0.4 |

**Table 1.**
*TEG materials and its performance [11].*

## 4.2 Properties influencing efficiency and output power

The measurement of the voltage generated by a material in response to a temperature gradient is commonly conducted using the Seebeck coefficient, which is also known as thermopower. An increase in the Seebeck coefficient results in a higher voltage output, hence offering potential benefits in enhancing the power output of the thermoelectric generator (TEG). The electrical conductivity of a thermoelectric material also exerts an influence on the flow of electric current via the thermoelectric generator (TEG). Enhanced electrical conduction and diminished electrical resistance are facilitated by heightened electrical conductivity, hence resulting in augmented power generation. The findings suggest that the utilisation of a solar thermoelectric generator featuring a well-thought-out thermal design can effectively optimise the advantageous characteristics of thermoelectric materials and substantially improve the efficiency of power generation [13]. In addition, a thermoelectric material's heat-transfer efficiency is reliant on its thermal conductivity. To enhance the thermoelectric efficiency and sustain a larger temperature gradient across the thermoelectric generator (TEG), it is imperative for thermoelectric materials to possess a diminished thermal conductivity. Utilising materials with restricted thermal conductivity prevents excessive heat transfer, hence facilitating a higher temperature gradient for power generation. Finally, it is worth noting that the figure of merit (ZT), a dimensionless parameter, encompasses the Seebeck coefficient, electrical conductivity, and thermal conductivity of a given material. A high ZT value serves as an indicator of strong thermoelectric performance. In thermoelectric generators (TEGs), materials that possess higher figures of merit (ZT values) generally exhibit enhanced efficiency and power generation capabilities. Tetrahedrites have the capability to generate a thermoelectric figure of merit (ZT) close to 1 at ambient temperature. However, recent advancements in the field have led to the synthesis and thermoelectric characterisation of a wide range of sulphide compounds, which exhibit remarkably high ZT values [12].

## 4.3 Advances in material research and development

Advancements in material research and development are contributing to the enhancement of thermoelectric generators (TEGs) in terms of their efficiency, power production, and cost-effectiveness. Researchers are currently doing extensive investigations into novel compounds that exhibit enhanced thermoelectric properties. This encompasses the processes of synthesising and evaluating novel chemicals, exploring alternative material categories, and enhancing existing materials through alloying, doping, and nanostructuring methodologies. There is a growing interest in the advancement of thermoelectric materials capable of operating at elevated temperatures, specifically over 600°C. The exceptional thermoelectric efficiency and high-temperature stability of these materials enhance the range of potential applications for TEGs. Subsequently, by employing nanostructuring methodologies, it becomes possible to modify the microstructure of thermoelectric materials at the nanoscale in order to enhance their performance. The reduction in grain size and the incorporation of nano-sized features can lead to a decrease in thermal conductivity while maintaining or even enhancing electrical conductivity. This leads to an augmentation in power generation and enhancement in thermoelectric efficiency. The concurrent objectives of decreasing heat conductivity and enhancing power factor in nanostructuring design methods have the potential to facilitate the advancement of nanostructured thermoelectric materials in future generations [14]. Furthermore, the advancement of contemporary manufacturing techniques has facilitated the production of intricate and highly efficient thermoelectric structures. Moreover, the implementation of material and geometrical optimisation techniques has the potential to enhance the thermoelectric generator's efficacy in augmenting the overall power generation of the system [15]. The advancement in thermoelectric module development is facilitated by techniques such as spark plasma sintering, additive manufacturing, and solution-based processing, which enable precise manipulation of material characteristics, hence enhancing performance. Finally, computer modelling and simulation approaches play a crucial role in facilitating the advancement and creation of novel thermoelectric materials. These approaches are employed to enhance the composition, structure, and doping levels of materials in order to optimise their thermoelectric performance. They aid in the discovery of potential candidates that possess desirable thermoelectric properties.

## 5. Design, construction, and operation of thermoelectric generators

### 5.1 System design considerations

When developing a thermoelectric generator (TEG), it is imperative to consider several critical variables. The choice of a heat source is contingent upon various factors, including the accessibility of surplus heat and the specific demands of the given application. Typical sources of heat encompass motor exhaust emissions, industrial furnaces, and concentrated solar power systems that harness solar energy. The presence of a substantial temperature differential is a crucial need for the functioning of the system [16]. The power output potential of a thermoelectric generator (TEG) is directly proportional to the magnitude of the temperature gradient between its heated and cold surfaces. The selection of the thermoelectric material is an additional critical determinant. The choice of materials is dictated by their thermoelectric characteristics, including the Seebeck coefficient, electrical conductivity, and thermal conductivity. Bismuth telluride (Bi2Te3) and its alloys are frequently employed as thermoelectric materials, alongside lead telluride (PbTe) and silicon-germanium (SiGe) alloys. The overall power output and system effectiveness are influenced by the configuration of the modules, which encompasses factors such as the number and layout of thermoelectric modules. Heat exchangers play a crucial role in facilitating efficient heat transfer between the heat source and thermoelectric generators (TEGs). These structures are specifically engineered to optimise the extent of contact with the surrounding surface area and are constructed using thermally conductive materials such as copper or aluminium.

### 5.2 Structural components and assembly

A thermoelectric generator comprises several essential components. Thermoelectric modules consist of a series of connection of thermoelectric devices. The composition of these elements involves the use of p-type and n-type thermoelectric materials that are placed between ceramic substrates [17]. The substrates serve the purpose of providing both electrical insulation and mechanical support, with the additional function of limiting heat transfer between the surfaces that are heated and those that are cool. The transport of heat from the heat source to the thermoelectric generator (TEG) is facilitated through the utilisation of heat exchangers. Heat transfer devices are commonly constructed with a substantial surface area in order to optimise the efficiency of heat transfer. These devices can assume various shapes, such as finned heat sinks, plate heat exchangers, and tube and shell arrangements. Adequate insulation is necessary in order to minimise heat dissipation from the thermoelectric generator (TEG) system, hence ensuring a larger temperature gradient for improved efficiency. Ceramic fibre, fibreglass, and aerogel represent insulation materials characterised by a diminished capacity for thermal conduction. The transmission of power from the thermoelectric generator (TEG) to the external load necessitates the utilisation of electrical connections. In order to minimise power losses, it is imperative that these connections have a low resistance. In general, metallic interconnects are commonly employed to facilitate the establishment of electrical connectivity between thermoelectric modules and external loads or power management systems.

### 5.3 Operation and control mechanisms

A variety of control systems are utilised in order to effectively manage the operation of a thermoelectric generator. Maintaining a consistent temperature differential across the thermoelectric generator (TEG) is crucial for effective temperature regulation [18]. The achievement of this outcome is facilitated by insulation, optimisation of heat exchange, and temperature monitoring devices. Control tactics, such as the implementation of feedback control circuits, can be effectively employed to manage the transfer of heat and uphold optimal operating conditions. Incorporation of power management systems within the TEG design enables the regulation of electrical output. These systems ensure compatibility with load or energy storage devices, encompassing voltage regulators, converters, and energy storage systems like batteries or capacitors. There are various approaches that can be employed to enhance the overall efficiency of the thermoelectric generator (TEG) system. These include optimising the thermoelectric materials, designing efficient heat exchangers, and integrating the TEG system with other energy conversion technologies. The proper functioning of the TEG necessitates the continuous monitoring of crucial operational variables, including temperature, voltage, and current. Monitoring systems typically comprise many components, including temperature sensors, voltage and current sensors, and control algorithms. Routine maintenance protocols, such as the washing of heat exchangers and conducting inspections, are necessary.

## 6. Applications of thermoelectric generators

Thermoelectric generators (TEGs) have diverse applications across various fields, offering efficient and sustainable energy solutions. This essay explores three prominent applications of TEGs: waste heat recovery and industrial applications, space exploration and satellite power, and remote power generation in off-grid locations.

### 6.1 Waste heat recovery and industrial applications

Thermoelectric generators (TEGs) play a vital role in the efficient use of residual heat, particularly within the context of industrial operations. A significant amount of residual heat is generated by several industries, such as manufacturing, electricity generation, and transportation. Thermoelectric generators (TEGs) have the capability to be integrated into exhaust systems or heat sources with the purpose of converting surplus heat into power that may be effectively utilised. Thermoelectric generators (TEGs) have the potential to be integrated into vehicle exhaust systems in order to harness the waste heat generated by the engine within the automotive sector [19]. The utilisation of this electrical energy has the potential to operate supplementary systems, leading to a reduction in the consumption of petroleum and an improvement in overall energy efficiency. Thermoelectric generators (TEGs) are also employed in several industrial applications, including metal casting and glass manufacture, to effectively harness and convert waste heat into electrical energy. The application of this strategy not only yields a reduction in energy consumption and greenhouse gas emissions but also offers significant cost benefits to enterprises.

### 6.2 Space exploration and satellite power

The utilisation of thermoelectric generators (TEGs) has been a critical component in the realm of space exploration missions and satellite operations, serving as a primary source of energy. Thermoelectric generators (TEGs) offer a dependable alternative in extraterrestrial environments, where traditional means of power generation may prove impractical. Radioisotope thermoelectric generators (RTGs) employ the process of radioactive decay, specifically using isotopes such as plutonium-238, to generate thermal energy and subsequently convert it into electrical power. The utilisation of Radioisotope Thermoelectric Generators (RTGs) in tandem with Thermoelectric Generators (TEGs) has facilitated the advancement of deep space expeditions, exemplified by the Voyager spacecraft. These missions persistently function and communicate valuable information from the furthest regions of our solar system [20]. Thermoelectric generators (TEGs) are employed in satellite applications to effectively convert surplus thermal energy generated by onboard systems into electrical power. This technology enhances the duration of satellite missions and diminishes the need for conventional battery power, hence ensuring uninterrupted operation of satellites in outer space.

### 6.3 Remote power generation in off-grid locations

Thermoelectric generators (TEGs) are a feasible solution for the generation of electrical power in distant regions without grid connectivity and facing limited access to electricity. Oftentimes, rural villages, remote research stations, and locations affected by disasters face challenges in accessing a reliable electrical infrastructure. Thermoelectric generators (TEGs) have the capability to produce electrical energy by harnessing heat sources that are easily accessible, including biomass burners, solar thermal collectors, and geothermal systems. The integration of thermoelectric generators (TEGs) with biomass burners has been found to offer multiple benefits in rural areas. These benefits include the provision of illumination, charging capabilities for small devices, and power supply for low-power appliances [6]. Solar thermal collectors and thermoelectric generators (TEGs) work in tandem to harness the ample solar energy available and convert it into electrical power. Similarly, thermoelectric generators (TEGs) have the capability to harness the thermal energy derived from geothermal systems located in locations with geothermal activity. TEG systems demonstrate a notable capacity for operation with little maintenance demands, thereby offering off-grid communities a viable and sustainable means of power generation. This, in turn, contributes to an improved quality of life for these people and facilitates the emergence of economic prospects [21].

## 7. Integration of thermoelectric generators into renewable energy systems

Thermoelectric generators (TEGs) can play a valuable role in the integration of renewable energy systems by converting waste heat into usable electrical power. They are solid-state devices that utilise the Seebeck effect to generate electricity when there is a temperature gradient across the device.

### 7.1 Challenge and opportunities

The incorporation of thermoelectric generators (TEGs) into renewable energy systems poses a range of obstacles and opportunities. The initial obstacles encountered pertain to efficiency. One of the primary obstacles encountered in the field of thermoelectric generators (TEGs) pertains to their comparatively lower efficiency in relation to alternative power generation technologies. At present, thermoelectric generators (TEGs) have a lower conversion efficiency compared to conventional technologies such as solar panels or wind turbines. Enhancing the efficacy of thermoelectric materials and devices is of paramount importance in order to optimise energy conversion and enhance the competitiveness of thermoelectric generators (TEGs) [22].

One of the potential advantages linked to the incorporation of thermoelectric generators (TEGs) into renewable energy systems is the recovery of waste heat. Thermoelectric generators (TEGs) present a distinctive prospect for the retrieval of waste heat from diverse origins and its subsequent conversion into practical electrical energy. The implementation of waste heat recovery has the capacity to enhance the overall energy efficiency of renewable energy systems, hence mitigating energy wastage and promoting sustainability.

### 7.2 Synergies with solar and wind energy

Thermoelectric generators (TEGs) have the potential to establish beneficial relationships with solar and wind energy systems, thereby augmenting their collective efficiency and performance. This academic text explores the potential of integrating thermoelectric generators (TEGs) with solar and wind energy systems in hybrid systems. It examines the ways in which TEGs can complement and integrate with these renewable energy sources. Thermoelectric generators (TEGs) have the potential to be effectively incorporated into hybrid systems that synergistically combine renewable energy sources such as solar or wind power with waste heat recovery. Solar panels and wind turbines can generate power through the utilisation of renewable energy sources. However, it is important to note that these systems also generate surplus heat. The integration of thermoelectric generators (TEGs) into these systems enables the capture and conversion of waste heat into supplementary electrical energy, thereby augmenting the overall energy production and enhancing system efficiency [23]. Through the strategic utilisation of the complementary attributes of thermoelectric generators (TEGs), solar energy systems, and wind energy systems, the generation of renewable energy can be enhanced in terms of efficiency, reliability, and sustainability. The integration of these systems holds the potential to optimise energy generation, minimise wastage, and make a significant contribution towards a more environmentally sustainable future.

### 7.3 Hybrid systems and energy management

Hybrid energy systems, characterised by the integration of various renewable energy sources alongside potentially conventional sources, can derive advantages from the implementation of efficient energy management solutions. The optimisation of operation and performance of hybrid systems is heavily reliant on effective energy management. The contribution of energy management to the success of hybrid systems will now be discussed. One of the contributions is the optimisation of resources. The practise of energy management in hybrid systems entails the strategic allocation and optimisation of various energy sources, taking into consideration issues such as their availability, cost, and environmental impact [24]. Through the examination of real-time data and projections, energy management systems possess the capability to ascertain the optimal amalgamation of energy sources in order to satisfy the system's demand, while concurrently minimising expenses and maximising the utilisation of renewable energy resources.

## 8. Future perspectives and expansion of thermoelectric generator production

The future perspectives and expansion of thermoelectric generator (TEG) production hold significant potential as advancements continue to be made in materials, manufacturing processes, and system integration.

### 8.1 Research and development trends

The field of thermoelectric generators (TEGs) is experiencing ongoing advancements in research and development (R&D), motivated by the objective of enhancing efficiency, cost-effectiveness, and scalability. One of the prominent research and development trends observed in the thermoelectric generator (TEG) business is the use of innovative materials. Scientists are currently engaged in the investigation of novel materials and the enhancement of pre-existing ones in order to optimise the performance of thermoelectric generators (TEGs). This encompasses the advancement of thermoelectric materials exhibiting enhanced thermoelectric properties, including elevated ZT values, reduced thermal conductivity, and heightened stability. Researchers are now investigating several material synthesis processes, including nanostructuring, doping, and composite production, in order to enhance the thermoelectric capabilities. The research and development trends are propelling innovation and expanding the limits of thermoelectric generator (TEG) technology. Researchers are actively engaged in overcoming obstacles and investigating novel opportunities to fully harness the capabilities of thermoelectric generators (TEGs) across diverse domains. These efforts are crucial in advancing the progress of sustainable and high-performance energy systems.

### 8.2 Emerging technologies and innovations

The advancements in thermoelectric generators (TEGs) are significantly influencing the trajectory of this technology, presenting novel prospects and applications. The following discourse highlights several noteworthy new technologies and advances within the thermoelectric generator (TEG) business. Flexible and wearable thermoelectric generators (TEGs) are a promising and rising technological advancement. The advancement of flexible and wearable thermoelectric generators (TEGs) has facilitated the incorporation of thermoelectric technology into several applications, including wearable electronics, smart clothing, and flexible gadgets. Furthermore, the utilisation of nanostructured and thin-film thermoelectric generators (TEGs) is also worth considering. Nanostructured and thin-film thermoelectric generators (TEGs) utilise advanced nanoscale engineering methodologies to improve their thermoelectric efficiency. Scholars are currently investigating the production of nanostructured thermoelectric materials, including nanowires, nanotubes, and thin films, with the aim of enhancing energy conversion efficiency. These technological improvements present the possibility of achieving elevated power densities, enhanced flexibility, and seamless integration with many other electronic equipment [25].

### 8.3 Scaling up production and commercialisation

The process of expanding manufacturing and facilitating commercialisation plays a pivotal role in the progression and extensive acceptance of thermoelectric generators (TEGs). The focus of scaling up the production and commercialisation of TEG is in the optimisation of manufacturing processes. Efficiently optimising manufacturing processes is necessary in order to facilitate the expansion of thermoelectric generator (TEG) output. This entails the enhancement of material synthesis, device manufacturing, and module assembly processes in order to enhance production efficiency and mitigate costs. The implementation of automation, quality control measures, and standardisation in the manufacturing process has the potential to enhance consistency and reliability, leading to a reduction in production time and expenses.

The process of expanding thermoelectric generator (TEG) production and bringing it into the commercial market necessitates a comprehensive strategy that spans various aspects. These include optimising manufacturing processes, reducing costs, ensuring quality control, doing market analysis, obtaining necessary certifications, implementing pilot projects, developing appropriate business models, and educating potential customers. By carefully addressing these elements, TEG technologies have the potential to penetrate wider markets, attain cost competitiveness, and make significant contributions towards the transition to a more sustainable energy future.

## 9. Conclusion

### 9.1 Summary of key points discussed

Thermoelectric generators, known for their advantageous characteristics such as simplicity, reliability, and environmental sustainability [26], offer a feasible alternative for converting thermal energy into electrical energy. The advancements in thermoelectric materials, comprehension of operational principles, variables affecting performance, and other related factors have facilitated the development of efficient generator designs. The versatility and potential impact of these generators are exemplified through their application in waste heat recovery, space exploration, and off-grid power generation. The integration of thermoelectric generators into renewable energy systems has the potential to contribute to a more sustainable energy mix. This approach also has opportunities for synergistic effects with solar and wind energy sources [27]. Future research and development endeavours should prioritise the exploration of advanced technologies, the improvement of material properties, and the expansion of production capabilities. These efforts aim to fully exploit the potential of thermoelectric generators and expedite their integration into various industrial sectors. Thermoelectric generators possess the capacity to make a substantial contribution towards meeting energy requirements while concurrently mitigating environmental impact through continued advancements.

### 9.2 Implications for sustainable energy solutions

The study of thermoelectricity holds significant implications for the development of sustainable energy solutions. Thermoelectric materials possess the capacity to substantially enhance energy efficiency across diverse domains, including industrial operations, automobile mechanisms, and power generation, through the retrieval and conversion of waste heat into usable electrical energy [28]. By reducing the release of greenhouse gases, this approach not only mitigates energy inefficiency but also fosters ecological sustainability. The utilisation of abundant waste heat sources, such as those found in industrial processes and power plants, to generate localised electricity is facilitated through the implementation of waste heat recovery systems employing thermoelectric devices. The utilisation of a decentralised method in this context serves to mitigate transmission losses and bolster the resilience of energy systems. Furthermore, thermoelectric technologies are utilised in distant and off-grid regions, offering a dependable source of electricity in situations when traditional power infrastructure is constrained. The integration of thermoelectric solutions with other renewable energy sources, such as solar and wind, has the potential to improve the stability and dependability of renewable energy networks [29]. The consequences underscore the capacity of thermoelectric research to propel the development of sustainable energy systems and foster a more effective and ecologically conscious approach to power generation.

### 9.3 Prospects and recommendations for further research

The exploration of potential avenues for future thermoelectric research and the formulation of suggestions hold significant potential for the advancement of sustainable energy solutions. The investigation of developing thermoelectric materials with enhanced performance, reliability, and efficiency is a critical field of research. In order to enhance the thermoelectric properties and overall efficiency, further research should be conducted to explore novel material compositions, nanostructuring techniques, and enhanced manufacturing procedures [30]. A thorough investigation is also necessary for the optimisation of thermoelectric devices and systems, encompassing the advancement of efficient heat exchangers and techniques for thermal management and integration. In order to explore novel opportunities and gain a deeper understanding of the fundamental principles governing thermoelectric phenomena, it is necessary to conduct research in the field of physics and thermodynamics. In the realm of thermoelectric research, it is advisable for future investigations to prioritise the advancement of thermoelectric materials and technologies. This emphasis is crucial in order to augment energy efficiency, optimise power generation, and facilitate sustainable energy harvesting for Internet of Things (IoT) sensor applications. By doing so, these efforts will not only contribute to energy conservation but also align with the objectives of achieving carbon neutrality and zero emissions [31]. In order to facilitate the progress of thermoelectric research, it is imperative to establish collaborative relationships among experts from several disciplines, including material science, physics, engineering, and computer science. Furthermore, it is imperative to do research on the usability, cost, and environmental impact of thermoelectric technologies in order to facilitate their widespread and pragmatic implementation. The domain of thermoelectric research has the potential to unlock the complete capacity of waste heat recovery and facilitate the development of efficient and sustainable energy solutions through the exploration of these specific research avenues.

## Author details

Baljit Singh Bhathal Singh
School of Mechanical Engineering, College of Engineering, Universiti Teknologi MARA, Malaysia

*Address all correspondence to: baljit@uitm.edu.my

## References

[1] Min G, Rowe DM, Zhou M. Recent advances in thermoelectric materials and systems: From material design to device integration. Progress in Materials Science. 2017;**87**:219-292

[2] Sottosanti L, Feser JP, Zhao Y. Thermoelectric energy conversion: Mechanisms, materials, and opportunities. Chemical Reviews. 2019;**119**(13):8311-8340

[3] Zebarjadi M, Esfarjani K, Dresselhaus MS. Perspectives on thermoelectrics: From fundamentals to device applications. Energy & Environmental Science. 2018;**11**(3):739-749

[4] Zhao LD, Chang C, Tan G, Kanatzidis MG, Snyder GJ. Anisotropic thermoelectric properties in layered copper chalcogenides. Nature Communications. 2016;**7**:13513

[5] Mallick R, Singh RS, Chen G, Ren Z. Nanostructured thermoelectric materials: Current research and prospects. Science. 2017;**357**(6358):eaak9997

[6] Singh B, Baharin NA, Remeli MF, Oberoi A, Date A, Akbarzadeh A. Experimental analysis of thermoelectric heat exchanger for power generation from salinity gradient solar pond using low-grade heat. Journal of Electronic Materials. 2017;**46**(5):2854-2859. DOI: 10.1007/s11664-016-5009-0

[7] Tundee S, Srihajong N, Charmongkolpradit S. Electric power generation from solar pond using combination of thermosyphon and thermoelectric modules. Energy Procedia. 2014;**48**:453-463. DOI: 10.1016/j.egypro.2014.02.054

[8] Cai Y, Xiao J, Zhao W, Tang X, Zhang Q. A general model for the electric power and energy efficiency of a solar thermoelectric generator. Journal of Electronic Materials. 2011;**40**(5). DOI: 10.1007/s11664-011-1616-y

[9] Lesage FJ, Pagé-Potvin N. Experimental analysis of peak power output of a thermoelectric liquid-to-liquid generator under an increasing electrical load resistance. Energy Conversion and Management. 2013;**66**. DOI: 10.1016/j.enconman.2012.10.001

[10] Karabetoglu S, Sisman A, Fatih Ozturk Z, Sahin T. Characterization of a thermoelectric generator at low temperatures. Energy Conversion and Management. 2012;**62**. DOI: 10.1016/j.enconman.2012.04.005

[11] Ding LC, Akbarzadeh A, Tan L. A review of power generation with thermoelectric system and its alternative with solar ponds. Renewable and Sustainable Energy Reviews. 2018;**81**:799-812. DOI: 10.1016/j.rser.2017.08.010

[12] Caballero-Calero O, Ares JR, Martín-González M. Environmentally friendly thermoelectric materials: High performance from inorganic components with low toxicity and abundance in the earth. Advanced Sustainable Systems. 2021;**5**(11). DOI: 10.1002/adsu.202100095

[13] Xiao J, Yang T, Li P, Zhai P, Zhang Q. Thermal design and management for performance optimization of solar thermoelectric generator. Applied Energy. 2012;**93**. DOI: 10.1016/j.apenergy.2011.06.006

[14] Vargiamidis V, Neophytou N. Hierarchical nanostructuring approaches for thermoelectric materials

with high power factors. Physical Review B. 2019;**99**(4). DOI: 10.1103/PhysRevB.99.045405

[15] Mahmoudinezhad S, Ahmadi Atouei S, Cotfas PA, Cotfas DT, Rosendahl LA, Rezania A. Experimental and numerical study on the transient behavior of multi-junction solar cell-thermoelectric generator hybrid system. Energy Conversion and Management. 2019;**184**. DOI: 10.1016/j.enconman.2019.01.081

[16] Li P, Cai L, Zhai P, Tang X, Zhang Q, Niino M. Design of a concentration solar thermoelectric generator. Journal of Electronic Materials. 2010;**39**(9):1522-1530. DOI: 10.1007/s11664-010-1279-0

[17] Abd Jalil MI, Sampe J. Experimental investigation of thermoelectric generator modules with different technique of cooling system. American Journal of Engineering and Applied Sciences. 2013;**6**(1):1-7

[18] Dalala ZM. Energy harvesting using thermoelectric generators. In: 2016 IEEE International Energy Conference (ENERGYCON). IEEE; April 2016. pp. 1-6

[19] Kumar S, Heister SD, Xu X, Salvador JR, Meisner GP. Thermoelectric generators for automotive waste heat recovery systems part I: Numerical modeling and baseline model analysis. Journal of Electronic Materials. 2013;**42**(4):665-674. DOI: 10.1007/s11664-013-2471-9

[20] Zoui MA, Bentouba S, Stocholm JG, Bourouis M. A review on thermoelectric generators: Progress and applications. Energies. 2020;**13**(14). DOI: 10.3390/en13143606

[21] Gao HB, Huang GH, Li HJ, Qu ZG, Zhang YJ. Development of stove-powered thermoelectric generators: A review. Applied Thermal Engineering. 2016;**96**:297-310. DOI: 10.1016/j.applthermaleng.2015.11.032

[22] Al Huneidi Dana I, Furqan T, Al-Ghamdi SG. Energy modeling and photovoltaics integration as a mitigation measure for climate change impacts on energy demand. Energy Reports. 2022;**8**:166-171

[23] Quan R, Li T, Yue Y, Chang Y, Tan B. Experimental study on a TE generator for industrial waste heat recovery based on a hexagonal heat exchanger. Energies. 2020;**13**(12):3137

[24] Reddick C, Sorin M, Bonhivers JC, Laperle D. Waste heat and renewable energy integration in buildings. Energy and Buildings. 2020;**211**:109803

[25] Ltd CDIP. Thermoelectric Generators (TEG) Market Projected to Expand at a CAGR of 12.64% from 2023 to 2030, According to Contrive Datum Insights [Internet]. GlobeNewswire News Room; 2023. Available from: https://www.globenewswire.com/en/news-release/2023/03/19/2629837/0/en/Thermoelectric-Generators-TEG-Market-Projected-to-Expand-at-a-CAGR-of-12-64-from-2023-to-2030-According-to-Contrive-Datum-Insights.html

[26] Aridi R, Faraj J, Ali S, Lemenand T, Khaled M. Thermoelectric power generators: State-of-the-art, heat recovery method, and challenges. Electricity. 2021;**2**(3):359-386. DOI: 10.3390/electricity2030022

[27] Wehbi Z, Taher R, Faraj J, Castelain C, Khaled M. Hybrid thermoelectric generators-renewable energy systems: A short review on recent developments. Energy Reports. 2022;**8**:1361-1370. DOI: 10.1016/j.egyr.2022.08.068

[28] Patyk A. Thermoelectrics: Impacts on the environment and sustainability. Journal of Electronic Materials. 2009;**39**(9):2023-2028. DOI: 10.1007/s11664-009-1013-y

[29] Strielkowski W, Civín L, Tarkhanova E, Tvaronavičienė M, Petrenko Y. Renewable energy in the sustainable development of electrical power sector: A review. Energies. 2021;**14**(24):8240. DOI: 10.3390/en14248240

[30] Chen Z, Han G, Yang L, Cheng L, Zou J. Nanostructured thermoelectric materials: Current research and future challenge. Progress in Natural Science: Materials International. 2012;**22**(6):535-549. DOI: 10.1016/j.pnsc.2012.11.011

[31] Mori T, Maignan A. Thermoelectric materials developments: Past, present, and future. Science and Technology of Advanced Materials. 2021;**22**(1):998-999. DOI: 10.1080/14686996.2021.1966242

Chapter 3

# Using Machine Learning Techniques to Discover Novel Thermoelectric Materials

*Ebrar Yildirim and Övgü Ceyda Yelgel*

## Abstract

Thermoelectric materials can be utilized to build devices that convert waste heat to power or vice versa. In the literature, the best-known thermoelectrics, however, are based on rare, costly or even hazardous materials, limiting their general usage. New types of effective thermoelectric materials are thus required to enable worldwide deployment. Although theoretical models of transport characteristics can aid in the creation of novel thermoelectrics, they are currently too computationally costly to be used simply for high-throughput screening of all conceivable candidates in the wide chemical space. Machine learning (ML) has been viewed as a promising technique to aid materials design/discovery because of its quick inference time. In this book chapter, we provide the whole workflow for machine learning applications to the identification of novel thermoelectric materials, predicting electrical and thermal transport properties and optimizing processes for materials and structures using cutting-edge ML methods.

**Keywords:** thermoelectric materials, machine learning, thermoelectric efficiency, electrical and thermal transport, thermoelectric figure of merit, rapid materials discovery

## 1. Introduction

The largest factor in what occurred to alternative energy sources was the rise in global energy consumption, which has now elevated to the status of a major societal issue. Thermoelectric (TE) technologies, with their ability to convert energy, will be crucial for the development of renewable resources in the near future. More than 60% of the released energy is lost as waste heat, so thermoelectric materials used for reducing or recovering waste heat are the most indispensable materials as an alternative energy source [1–3]. TE materials: according to the scientific phenomena known as the "Seebeck Effect," "Peltier Effect" and "Thomson Effect," either the electric potential causes the temperature difference, or the temperature difference causes the electric potential. Although the TE effect is not zero in any material, it is too low to be useful in the vast majority of materials. The TE effect ensures that almost absent or rarely used heat energy is used as efficiently as possible. The thermoelectric effect has several benefits, such as being reliable, scalable, quiet, portable and ecologically beneficial. There are a variety of applications for thermoelectric materials, including refrigeration, waste heat recovery, energy conversion systems, infrared sensors, space missions, and research [1–3].

The thermoelectric efficiency coefficient "$ZT$" a dimensionless value, indicates the performance of TE materials:

$$ZT = \frac{S^2\sigma}{\kappa_{\text{total}}}T \tag{1}$$

where $S$ is the Seebeck coefficient; T is the temperature and $\kappa_{\text{total}} = \kappa_{\text{ph}} + \kappa_{\text{c}} + \kappa_{\text{bp}}$ is the total thermal conductivity. Total thermal conductivity consists of three components: phonon doping $(\kappa_{\text{ph}})$, carrier doping (electron or hole) $(\kappa_{\text{c}})$ and bipolar doping (electron-hole pair) $(\kappa_{\text{bp}})$ [4]. Materials with high electrical conductivity and Seebeck coefficient and low overall thermal conductivity should be selected to have high TE performance and efficiency. Since these three physical quantities are related to physical laws (Boltzmann transport equation $S - \sigma$; Wiedemann-Franz law $\sigma - \kappa_{\text{c}}$), materials with these properties are very rare. Because of this, it is also rather difficult to determine these thermoelectric characteristics experimentally or theoretically.

The efficiency limit required for the widespread use of thermoelectric systems is to reach the ZT > 1 value, and thus, 10% conversion efficiency is achieved [5]. To address the current bottleneck of TE technology, rapid research and discovery of new TE materials with desired performance are required. Finding the best-performing thermoelectric material for various thermoelectric module layouts is the major problem, much like discovering many other energy technology potentials (such as solar panels, solid-state batteries and catalysts). Materials analysis and design have grown in popularity in recent times, with an emphasis on statistical modeling, development and the discovery of new materials with specific properties. Thanks to *ab initio* methods, the computation of material attribute bonds has been successful since the 1970s. It is possible to perform large-scale computations with the available computing power of supercomputer clusters. However, despite their enormous computational power, these processes still require hundreds or thousands of core years. Therefore, artificial intelligence (AI) technology appears to be one of the most promising and extremely important approaches in the discovery and design of next-generation materials. The data-driven approach phase of AI technology is when it learns from previous information to create prediction functions. In order to assist in the discovery, design and optimization of novel materials at this stage, data-driven techniques employ material knowledge from experimental data or high-throughput simulations. TE properties ($S$, $\sigma$, $\kappa_{\text{total}}$ and $ZT$) can be accurately predicted using artificial intelligence techniques. The combination of big data and machine learning algorithms will be able to explore the potential of existing research data at the highest level and develop the necessary methods for the emergence of next-generation TE materials. Unfortunately, the number of studies on machine learning methods focusing on next-generation TE material discovery in the literature is quite limited so far. It should also be noted that although artificial intelligence techniques are more effective than experiments and theoretical computational methods, they severely constrain the creation and growth of reliable predictive models due to a lack of data and skewed data. In addition to that materials science includes various data categories such as organic materials, metal, semiconductor materials, etc. [6]. Unlike AI-powered fields such as image or natural language processing, the outputs of materials science cannot be easily obtained and verified which makes it difficult to obtain valuable datasets [7–9]. Material innovations played an important role in the science and technology revolution. Consideration should be given to AI technology as one of the most promising and important strategies for the development of educational programmes and the next generation of educational institutions. Data-driven approach habits resulting from grasping the predictive perspective by learning from existing features are now the dominant artificial intelligence technologies. Data-driven techniques leverage materials knowledge from experimental data or high-throughput calculations to aid material procurement, design and optimization [10]. Even if they are more effective than experiments and theoretical methods, AI still has a long way to go. One of the most significant barriers to improving materials science using AI technologies is a lack of data. The construction and development of sophisticated prediction models using AI are severely hampered by inefficient, skewed data. AI approaches may reliably anticipate the TE characteristics utilized in the dimensionless thermoelectric value of merit (such as electrical conductivity, thermal conductivity, and/or Seebeck coefficient) [11–13]. Furthermore, few studies have been conducted to identify representative TE material descriptors and AI models for predicting TE properties [10].

Thanks to the endless contribution of the TE foundation, the enormous observation obtained together with the research data makes it possible to reach the data-driven approach. The development of AI approaches has tremendously aided in the better utilization of these resources. The combination of big data and machine learning algorithms will be capable of exploring the full potential of existing research data and developing the procedures required for the creation of next-generation TE materials. Data-driven methods and dominant artificial intelligence technologies are in the data-driven approach stage. It is widely used in materials science to learn from existing data, predict mechanical properties and responses, develop predictive functions, scan materials and facilitate design. Materials information from experimental data or high-output calculations is used in data-driven methodologies to improve materials discovery, design and optimization. TE properties such as electrical conductivity, thermal conductivity and Seebeck coefficients, which are also used in dimensionless thermoelectric value coefficients, can be reliably predicted using artificial intelligence approaches. On the other hand, the amount of machine learning techniques research focusing solely on TE materials is relatively limited. In addition, several studies have been conducted to identify representative TE material descriptors and AI models to predict TE properties [14–16].

The purpose of this review article is to provide a comprehensive overview of the most effective strategies for predicting properties and optimizing processes for materials and structures using cutting-edge ML methods and machine learning-driven optimization approaches in materials science with a focus on thermoelectric materials to make a presentation. Our paper begins with the thermoelectric material data-generating approach and then moves on to the use of machine learning models and a review of the literature. We believe that our paper's breadth is meant to be useful to both academia and industry as a reference for reaching thermoelectric materials research and advances. **Figure 1** shows the logistic structure of our article. In the first half of our article, we will discuss how a thermoelectric database can be generated from theoretical calculations or practical observations to describe new materials using machine learning approaches. In the second part of our article, machine learning model evaluation methods in TE materials and different machine learning techniques will be explained respectively, and which training model from machine learning

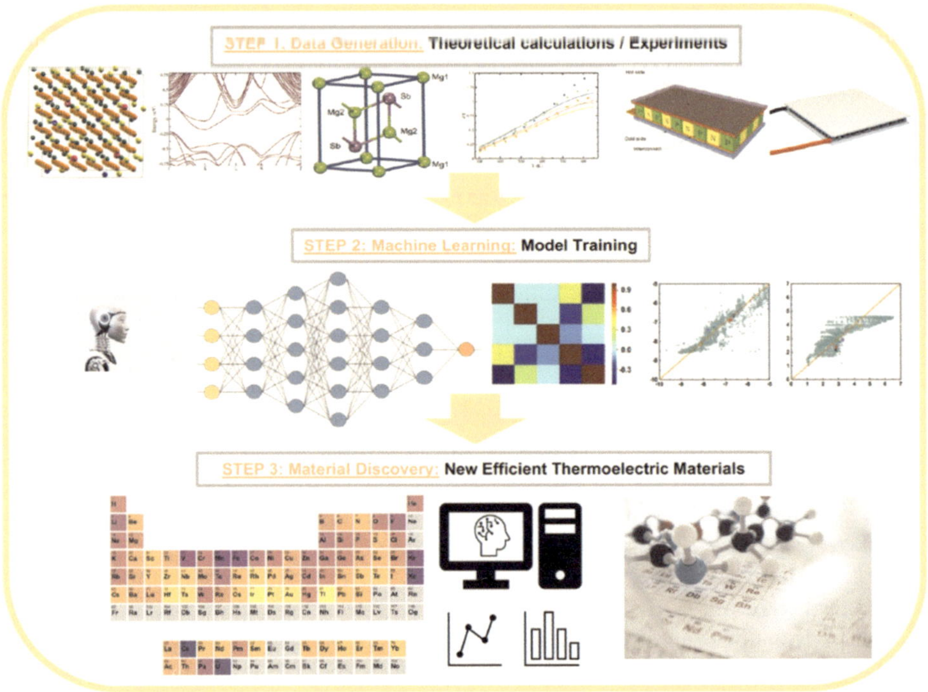


**Figure 1.**
*The proposed workflow of this chapter. The first step is learning from theoretical calculations or experimental results thus thermoelectric database can be created. The second step is machine learning model training and choosing the best suitable algorithms to make highly accurate predictions. The third step presents the discovery of new thermoelectric materials.*

methods will help to discover new materials with higher accuracy will be presented with a comparative discussion. The studies that have been done in the literature so far will be listed in detail in the last part of our paper.

## 2. Thermoelectric database generation

Machine learning-based approaches use materials knowledge from experimental data or high-throughput theoretical calculations to aid in new thermoelectric materials discovery, design and optimization. We can give these database sources and their explanations as follows.

### 2.1 From experimental characterizations

To learn models, machine learning needs a database of previously acquired knowledge. The materials database has to be marked with the appropriate TE attributes for machine learning-guided TE materials discovery to work. The TE efficiency dominates three combinatorial material properties: Seebeck coefficient, electrical conductivity and thermal conductivity. Because of the influence of these three combinatorial factors on each other, maximizing these three parameters is the most difficult challenge in TE materials. Thus, these three combinatorial parameters are the most popular tags for the TE machine-learning computer model. Phonon and electron

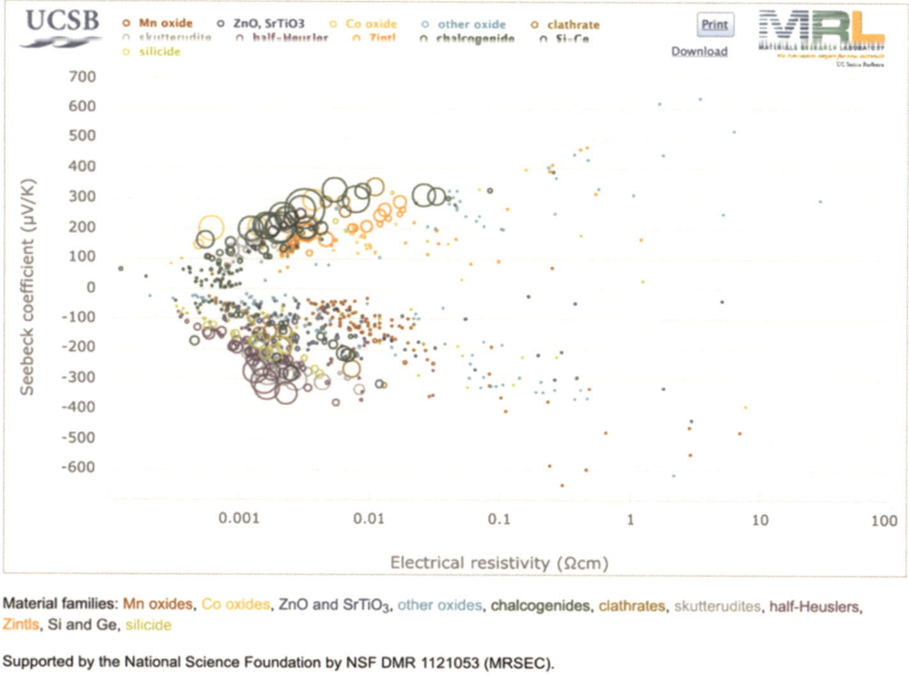


**Figure 2.**
*A snapshot of the UCSB website and a look at its contents.*

distribution are considered as the labels of choice from the controls, band structure, band gap and phonon distribution of TE materials. The creation of a machine learning-based material discovery tool requires access to vast volumes of data where the learning process can provide accurate correlations between input and output pairs. The most typical application of these TE labels for various materials is characterization. Tools for generating fundamental TE materials include machine learning models for the results of TE-related processes from diverse materials research fields. The UCSB database is one of the most extensive TE materials databases [17, 18]. The UCSB database includes various component TE properties for more than 1000 different users, abstracting information from more than 100 publications. As shown in **Figure 2**, the UCSB database combined with appropriate visualization tools can provide users with an efficient approach to developing new TE materials from experimental characterizations.

### 2.2 From theoretical calculations

The first-principles computation of atomic-scale materials is another method for obtaining TE data for materials, in addition to TE data through experimental characterization. When compared to the experimental characterization database, the theoretical calculation results have a standardization value, which implies that the TE efficiency results will not be influenced by equipment, human or measurement error. The Seebeck coefficient, carrier thermal conductivity, phonon thermal conductivity and electrical conductivity can be computed using the Boltzmann transport equation.

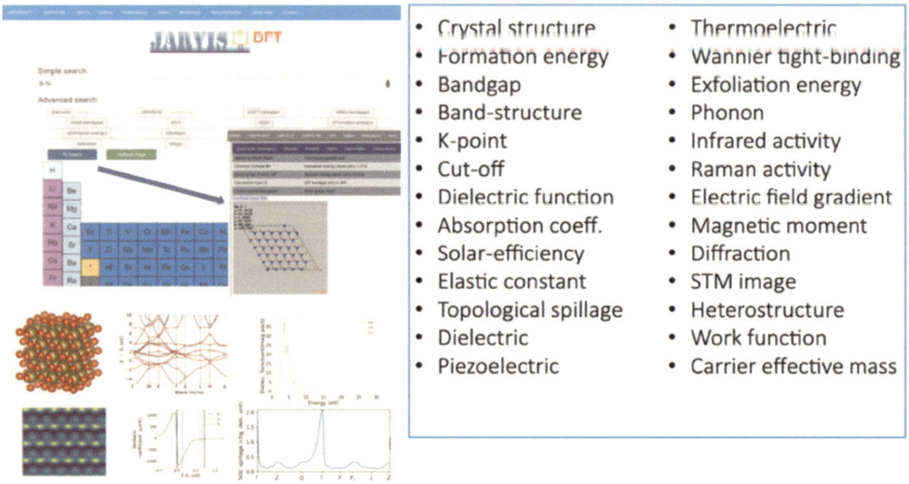


**Figure 3.**
*A snapshot of the JARVIS-DFT website and summary of its contents.*

The first-principle calculations equation is capable of producing trustworthy and accurate TE data, but it is computationally costly. As a result, it is challenging to fulfill the need for a huge amount of data for machine learning using first-principle computation. High-throughput first-principle computing was developed to tackle this dilemma [19]. The computational cost can be greatly decreased in high-throughput first-principle research with some accuracy loss [20–22]. High-throughput first-principle calculations can save results to massive material databases for later use, such as quick material scanning. As shown in **Figure 3**, the JARVIS-DFT database contains TE performance data of approximately 36,000 three-dimensional and 900 two-dimensional materials from density functional theory calculations (DFT). Along with the electronic thermal conductivity, electrical conductivity and Seebeck coefficient, the JARVIS-DFT cage also improves thermal conductivity. To create a machine learning classification model for prescreening materials with good TE characteristics, this data is also utilized. Additionally, **Table 1** shows a selection of a publicly available list of datasets of thermoelectric characteristics that may be utilized for machine learning.

## 3. Machine learning (ML)

ML terminology and all related definitions will be explained in this section and will help the readers better understand and become familiar with the various machine learning categories. As depicted in **Figure 4**, ML models are classified into three types: supervised, unsupervised, semi-supervised, and reinforcement learning (RL). Input and output variables, sometimes referred to as independent and dependent variables, are included in the training dataset for supervised learning. We may envision a dataset in the field of materials science that includes both chemical and physical attributes. Structures are the independent variables, whereas material attributes are the dependent variables. The machine learning algorithm is programmed to learn the function that represents the connection between independent and dependent variables.

| Dataset | Year | References | Data source | Compounds | Features |
|---|---|---|---|---|---|
| Wang et al. | 2011 | [23] | Theory | 2585 | PF, $m^*$ |
| Carrete et al. | 2014 | [24] | Theory | 450 | $\kappa_{ph}$ |
| TE Design Lab. | 2016 | [25, 26] | Theory | 2701 | $\kappa_{ph}$, μ, $m_D^*$ |
| Ricci et al. | 2017 | [27, 28] | Theory | 47,737 | $\sigma$, S, $\kappa_C$ |
| Xi et al. | 2018 | [8] | Theory | 161 | PF |
| Chen et al. | 2019 | [29] | Experiment | 100 | $\kappa_{ph}$ |
| Starrydata2 | 2019 | [30, 31] | Experiment | 434 | $\sigma$, S, $\kappa_{total}$ |
| Priya et al. | 2021 | [32, 33] | Experiment | 585 | $\lambda_i$ |
| Jaafreh et al. | 2021 | [34] | Theory | 119 | $\kappa_{ph}$ |
| Miyazaki et al. | 2021 | [35] | Theory | 143 | $\kappa_{ph}$ |
| MIP–3d | 2021 | [36, 37] | Theory | 4400 | $\sigma$, S |
| Tranås et al. | 2022 | [38] | Theory | 122 | $\kappa_{ph}$ |

*Here, the physical properties are indexed as follows: PF is the power factor,* $m^*$ *is the carrier effective mass,* $\kappa_{ph}$ *is the phonon thermal conductivity,* μ *is the carrier mobility,* $m_D^*$ *is the density of states effective mass, S is the Seebeck coefficient,* $\kappa_c$ *is the carrier thermal conductivity,* $\kappa_{total}$ *is the total thermal conductivity, and* $\lambda_i$ *is the ionic conductivity.*

**Table 1.**
*A list of publicly available datasets of thermoelectric properties that can be used for machine learning is presented.*

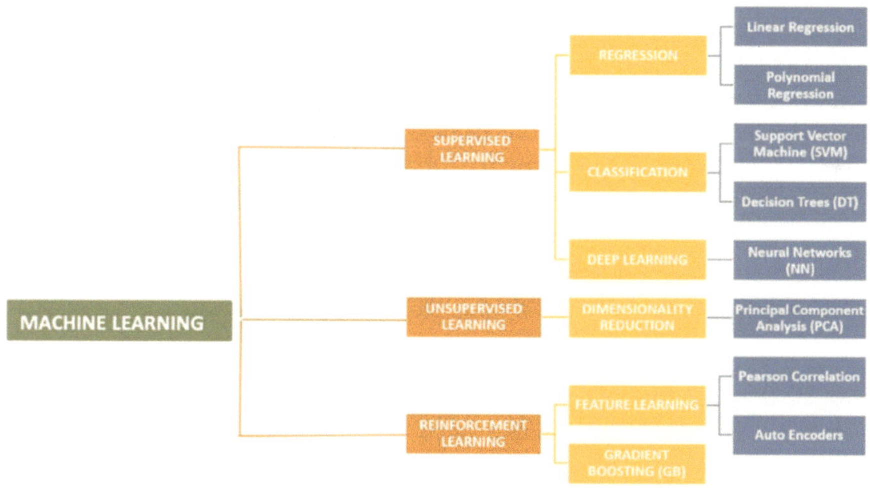


**Figure 4.**
*Schematic representation of machine learning algorithms. The hierarchy of several machine learning algorithms, including supervised, unsupervised and reinforcement learning methods.*

## 3.1 Supervised learning

The use of labeled datasets to train algorithms for reliable data classification or result prediction characterizes supervised learning. The structure of process steps in supervised learning is shown in **Figure 5**.

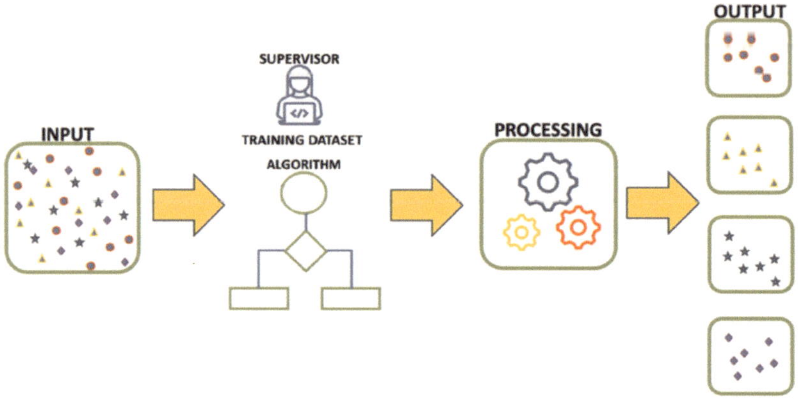


**Figure 5.**
*Supervised learning workflow.*

### *3.1.1 Regression*

Most of the work on the use of machine learning models in the TE material field has been done with regression. Regression models are models that consistently give consistent values when input is given. The link between a single or several quantitative or categorical independent variables and categorical quantitative dependent variables is represented by regression models. To determine the Seebeck coefficient, electrical conductivity, thermal conductivity, etc., in the study of TE materials, researchers need well-organized and well-chosen material characteristics. It uses sophisticated regression frameworks to forecast desirable material characteristics such regression models may be roughly categorized into two groups: deep learning models, which primarily rely on neural network theories, and classical statistical learning models, which are typically based on classical statistical learning theories. Support vector regression, tree-based models, Gaussian processes, and linear regression are the statistical learning techniques that are most frequently utilized in research that already exist to predict the characteristics of TE materials.

### *3.1.2 Linear regression/multiple regression (LR/MR)*

The linear modeling of a scalar response's connection with one or more explanatory variables, also referred to as the dependent and independent variables, is known as linear/multiple regression [39]. Simple linear regression refers to a scenario where there is only one variable, while multiple linear regression refers to a situation when there are numerous variables. This varies from the phrase "multivariate linear regression," which predicts numerous linked dependent variables rather than a single scalar variable. The assumption of a linear regression model is that the regression function will be linear in terms of the input variables $X_1, \ldots ,X_2$, which may be descriptors of the input material. The fundamental linear regression model corresponds to this [39],

$$f(x) = \beta_0 + \sum_{j=1}^{p} X_j\beta_j. \tag{2}$$

In this case, $\beta_0$ stands for the learned bias term and $\beta_j$ for the learned weight corresponding to $X_j$. Between the input independent variables and the output dependent variable, linear models presuppose a linear connection or a good approximation of a linear relationship. In this paradigm, the input variables might be the original quantitative or categorical values, such as material descriptors or transformations of the original values, like log, square root, polynomials or other transformations. The parameters in Eq. (2) are estimated with ordinary least squares (OLS) minimizing the squared error shown as [39],

$$\min_{\beta} \sum_{i=1}^{N} (y_i - f(x_i))^2 \tag{3}$$

where N is the number of samples in the training set, $x_i$, i. for example, is the feature vector, and $y_i$, i. is the sample's actual value. Minimizing Eq. (2) to estimate parameters is equivalent to solving a normal equation as [39]

$$\hat{\beta} = (X^T X^{-1}) X^T y \tag{4}$$

where X is the feature matrix and Y is the actual target vector. While the basic form of linear regression is simple and useful in many scenarios, it can lead to overspread if the learning for some variables is too large to be the consumer, but if the training observes the observations but not the unseen thought. Contraction strategies can be employed to alleviate the model's significant variability. The most commonly used regularizations are ridge editing and Lasso editing. Both ridge and Lasso regression limit the size of the parameters for oversteer prevention entering the OLS, the penalty terms controlled by the amount λ [39].

In ridge regression, the model encourages small parameter size using the L2 norm, while in Lasso regression, the model encourages a value of 0 using the L1 norm. When the feature matrix is invertible, it is possible to estimate the parameters in each of the three models by solving linear systems. A different approach is to use gradient-based optimization methods like stochastic gradient descent to minimize f(x) regardless of whether the feature matrix is reversible as [39]

$$\min_{\beta} \sum_{i=1}^{N} (y_i - f(x_i))^2 + \lambda \sum_{i=1}^{N} \beta^2{}_j \tag{5}$$

$$\min_{\beta} \sum_{i=1}^{N} (y_i - f(x_i))^2 + \lambda \sum_{i=1}^{N} \beta_j. \tag{6}$$

Material descriptors and their transformations can be handled as input independent variables and desired material qualities as output dependent variables when linear regression models are applied to TE materials. In order to evaluate the power factors of sintered powders, Wang et al. used a linear regression analysis [23]. They found that the power factor was strongly correlated with the electronic band gap and carrier effective mass. Utilizing PCA-transformed features, Reokeghem et al. [40] used the linear regression model to calculate the force constants of semiconductor oxides and fluorides with cubic perovskite structures at various temperatures. To calculate the elastic bulk and shear modulus of polycrystalline materials, De Jong et al. built a polynomial feature basis using composition and structural descriptors and used Lasso regression with gradient boosting [41]. The trained model was also utilized to scan very hard materials. Miller et al. looked at the use of the linear regression model, along with other techniques, to forecast the carrier concentration range of semiconductors that resemble diamonds [42]. Given the composition of ionic radii, Li et al. utilized Kernel Ridge Regression to predict the dissociation energy and verified the training model using the formability of actual perovskites [43]. The regression model's success demonstrated that the experimental engineering of stable perovskites might be guided by machine learning techniques when applied to DFT-computed data. Iwasaki et al. [44] calculated thermal power using quadratic polynomial lasso regression and elastic mesh. When introducing machine learning methods to a particular regression problem for the first time, researchers should take into account linear regression models since they are straightforward and simple to comprehend. However, only a few cases allow for the validity of the assumption that there is a linear relationship between the input features and the output target. For TE materials with complex nonlinear interactions between input descriptors and material properties, models that can capture nonlinearity should be taken into consideration.

### *3.1.3 Classification*

#### *3.1.3.1 Support vector machine (SVM)*

Support vector networks and support vector machines (SVMs) are supervised learning models for data prediction. As a non-distributed binary output classifier, SVM training technology develops a model that redistributes one of the two categories in accordance with prior institutions [45]. In order to increase the separation between the two groups, DVM carefully maps training samples to points. New birds are calculated and mapped to the same region based on which home the inhabitants are from. By incorporating them into high-dimensional feature areas that they enter, DVMs employ the use method to build non-persistent environments. The same theory as SVM underlies Support Vector Regression (SVR), a regression method. When data is in an unlabelled format, supervised learning rights are available. Therefore, it incorporates an unsupervised learning method where fresh data is mapped and data is organically sorted into categories. Support vector clustering algorithms classify unlabelled data by using support vector statistics of animal motions from SVM. It is one of the clustering jobs for industrial computers that is most frequently employed [45].

#### *3.1.3.2 Decision trees (DT)*

Decision trees, also known as tree-based models, are not parametric and as a result, this method is employed in supervised learning for both classification and regression. A decision-making tool called DT makes use of a tree-based representation of options and possible consequences [46]. By nesting the data on certain property values depending on specified parameters, they may capture nonlinear correlations between the predictive and target variables. The goal of each split is to generate more homogenous datasets in which the target values are more comparable to each other than they were before the split. Decision trees perform a search that covers the entire dataset, find the feature and the split value, then observe each distinct value of each feature, and perform the task of dividing the data into two subsets. As a result, total errors are minimized. Thus, the conditions used for data splitting are determined by the potential homogeneity of the target values. Decisions are made in the leaves of the algorithm as the data is divided into nodes. In classification trees, the decision variable is categorical [47]. Due to this mechanism, tree models can be used to determine the importance of a feature, as the most effective feature will be the one to split the data first. One of the biggest problems decision trees have is overfitting. A single tree tends to overfit the training set (it is sensitive to changes in the training sets), leading to over-learning on previously unknown data. To address this problem, ensemble methods such as random forests [48] and gradient-enhancing trees [49] can be used. There are several advantages to using linear and non-parametric tree-based regression models. First, tree-based models provide great interpretability since decisions are made in a certain sequence based on features and their values. Second, tree-based models can handle both categorical and continuous input characteristics intuitively, with no data preprocessing required. Lastly, while automatically reflecting the significance of input characteristics, tree-based models can capture the complex nonlinear relationships that exist between input-output pairings. In order to estimate TE material qualities from material descriptions, tree-based models are frequently utilized.

Carret et al. used random forest regression to predict the lattice thermal conductivity for semi-Heusler compounds based on chemical, compound, and particular thermal conductivity information in some of the decision tree investigations in materials discovery [24]. The learned regression model was used to examine the thermodynamic stability as well. According to Gautois et al., they trained a random forest model to estimate the Seebeck coefficient, thermal conductivity, electrical resistance, and band gap using data from the periodic table [50]. Additionally, the developed model successfully suggested a unique molecule from the real chemical space that could be tested experimentally, demonstrating the potential of employing machine learning techniques to guide materials discovery and design.

Furmanchuk et al. [51] utilized the random forest to quickly anticipate the properties of materials that were experimentally synthesized and to determine the Seebeck coefficient of crystalline materials. Miller et al. employed the random forest in addition to the linear model to calculate the repeatability and range of carrier concentrations for semiconductors that resemble diamonds [52]. For this objective, the random forest did not, however, perform better than the linear model. In order to estimate the interface thermal resistance between two materials using well-selected physical, chemical and material attribute descriptors, Wu et al. employed LSBoost's regression tree assemblages [52]. With all descriptors, the ensemble model's coefficient of determination (R2) was 0.919, while with just feature descriptors and thickness, it was 0.907. Iwasaki et al. employed a decision tree regression model in addition to an elastic mesh and quadratic polynomial Lasso regression inside a linear model framework to estimate thermopower [53]. The model assisted in investigating the underlying physics of the spin-driven TE phenomenon and in developing materials that exhibited these effects. The ability of the random forest regression model to forecast the success rate (ZT) of hot extruded $Cu_xBi_2Te_{2.85+y}Se_{0.15}$ TE materials was studied by Wang et al. [53].

### *3.1.4 Deep learning*

Deep learning (DL) approaches are based on neural network theories, which vary from traditional machine learning techniques in that processes can be represented by linked neurons [54]. Artificial neural networks (ANNs) are suitable estimators for every function, according to the universal approximation theorem [55]. Multiple connected layers of neurons might solve the intractability issue while retaining performance instead of having a high number of neurons in a single layer to capture complicated mapping within the data. Deep learning techniques may learn many layers of representations of the original input data, which are created by nonlinear modules modifying the representation one level at a time [56]. Deep learning techniques can also capture complex mappings. The capacity for representation learning permits the use of the most unprocessed material and does not need extensive feature engineering or selection. The feed-forward fully connected neural networks (FCNN) are mostly employed in the application of TE materials. Despite having advantages over traditional machine learning models, deep learning is sometimes over-parameterized and hence needs a lot of data to be able to acquire a good mapping that generalizes effectively. Due to data scarcity and data sparsity, this imposes a significant limitation when using deep learning algorithms to predict material attributes. In addition, the lack of a pre-defined model shape and the complicated hierarchy of layers and neuron activation make deep learning models difficult to explain.

#### *3.1.4.1 Neural networks (NN)*

The design and operation of biological neural networks served as the inspiration for the machine learning technique known as neural networks. They are made up of layers of linked nodes that process incoming data and generate output. To recognize pictures, comprehend spoken language and forecast time series, among many other tasks, neural networks are utilized [57–59]. As learning advances, the weights of connections, which are referred to as edges, alter. By using different layers, different adjustments to inputs are carried out. A known input, a known output, and probability-weighted associations that are recorded in the network's data structure are used to train ANNs. To train a neural network, one uses the error, which is the discrepancy between the processed output of the network and the desired output. According to a learning rule and error value, the network's weighted associations are updated. When a certain number of modifications yield outcomes that are somewhat near to the expected outcomes, training may be considered complete. A wider family of machine learning algorithms built on representational learning and neural networks includes DL. When extracting higher-level characteristics from raw data using several layers, DL is a form of machine learning methodology that is more accurate than other machine learning techniques. Examples of DL methods include deep neural networks (DNNs), convolutional neural networks (CNN), recurrent neural networks (RNN), ANN, and convolutional neural networks (RNN). Hidden layers allow data to go from the input layer to the output layer. Despite the fact that DL approaches are more accurate and effective than other machine learning techniques, they still need a lot of data and are computationally costly because of the numerous parameters that must be optimized during training.

### 3.2 Unsupervised learning

Unlabelled datasets are analyzed and clustered using machine learning techniques in unsupervised learning. Without requiring human participation, these algorithms identify occult patterns or data clusters. In **Figure 6,** the structure of process steps in unsupervised learning is shown.

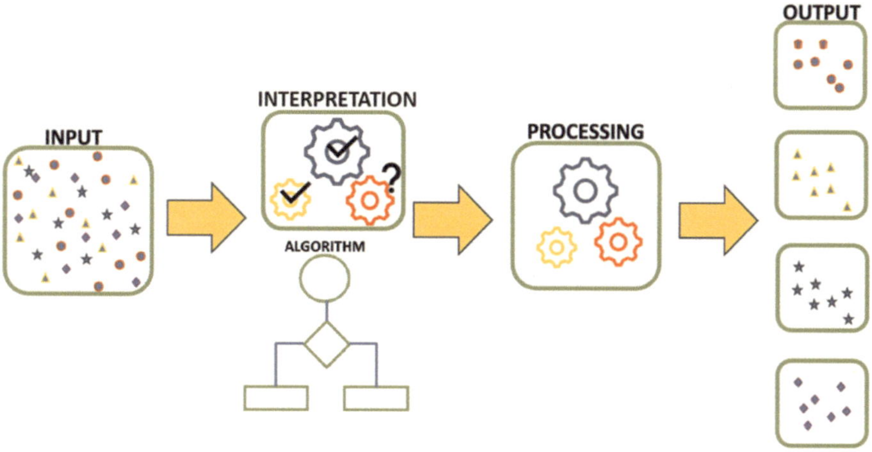


**Figure 6.**
*Unsupervised learning workflow.*

### *3.2.1 Principal component analysis (PCA)*

A typical linear dimension reduction approach used to extract significant information from various datasets, converting the input features to a new coordinate to minimize the number of features while keeping the majority of the original information [60]. It provides a roadmap on how we can reduce complex datasets to a smaller size to reveal a simplified structure. Because of its simplicity, it is a fundamental analysis used in data analysis and other domains. To produce the principle components whose data captures as much variation as feasible, linear combinations of the original attributes are used.

A certain way is followed to establish the basis. The first fundamental explanation is the linear approximation of the original features with the greatest variability among all possible combinations. Linear operations of the original features that contain the most variability among the remaining components are considered second. The other component continues by seeing this rule. Therefore, the basic configuration of a dataset can provide the best-going approaches. PCA execution procedures are listed below.

- Data are standardized.
- Construct a covariance matrix for self-division.
- Extract the eigenvalues of the eigenvectors of the covariance matrix to determine the basis determinations.
- It is tried not to be accepted as the optimum main.

In TE material machine learning studies, PCA has been applied in most applications to reduce the input dimensions during model creation [40]. If we give an example of applications where PCA is used. In their study of estimating force constants, Roekeghem et al. used PCA to transform the original descriptors and selected the top 10 principal components as regression model input [40]. Wagner and

Rondinelli [61] employed PCA to convert strongly correlated mode characteristics and the first three principal components in conjunction with decision trees to forecast high-temperature perovskites.

### 3.3 Reinforcement learning

Reinforcement learning is a machine learning training method based on rewarding desired behaviors and/or punishing undesired ones, as displayed in **Figure 7**. In general, a reinforcement learning agent can perceive and interpret its environment, take actions and learn through trial and error.

#### *3.3.1 Gradient boosting (GB)*

A class of supervised machine learning techniques known as gradient boosting employs a group of weak learners to produce a strong learner. The approach works by gradually introducing weak learners into the group, each of whom corrects the mistakes of the one before them. Decision trees are frequently used as weak learners, and a gradient descent technique is used to train the ensemble. As a result, the model is more precise than any of the individual subpar students. XGBoost, LightGBM, and CatBoost are a few well-known gradient-boosting algorithms [62]. Compared to other machine learning methods like random forests and support vector machines, gradient boosting approaches often offer superior accuracy. This is so that the algorithm may gradually increase its accuracy by learning from the errors of earlier, less accurate learners.

#### *3.3.2 Feature learning*

It is crucial to use appropriate descriptors that provide sufficient details for the associated attributes if one wants to forecast material properties effectively and precisely. It is unclear how to pick the appropriate descriptors from the vast array

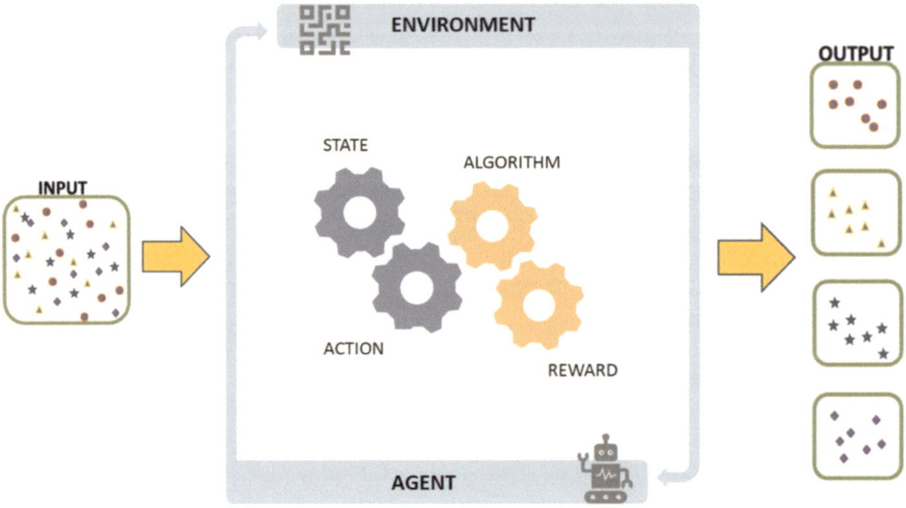


**Figure 7.**
*Reinforcement learning workflow.*

available to characterize the nature of materials. Additionally, in order to prevent overfitting and ensure that they generalize adequately to new, unknown data, machine learning algorithms often require a substantially higher number of samples than the number of characteristics or identifiers. To address this issue, efforts have been made to choose descriptors that are instructive or to transform the descriptors into a feature set with a reduced dimension while retaining the original data. These two strategies are called feature engineering and feature selection, respectively [63]. Without domain knowledge, descriptors can be derived or mixed based on competence in data-driven feature selection and engineering. Instead, supervised or unsupervised methods can be used to do feature selection and engineering. Currently, Pearson correlation, principal component analysis, and automated encoders are the most widely used techniques for data-driven feature selection and engineering for TE materials.

The foundation for the effective use of machine learning algorithms in the design and discovery of TE materials is the search, identification and selection of relevant, dominating material descriptors or features with enough numerical weight to enable precise model predictions. It is common to refer to feature fingerprints or identifiers as the collection of arguments that must be provided into a certain model. A strong awareness of hidden relationships between input and desired output as well as domain expertise, are frequently required for feature selection. Problems emerge when important material identifiers are missing from the original dataset or when feature engineering of particular inputs is not possible in order to numerically express these descriptors. The choice of features should be made so that the influence of each individual input variable on the final dependent target output is significant but not always clear. In other words, one's intuition, skill or subject knowledge plays a major role in successfully finding and choosing relevant qualities. The use of intuition in feature engineering is optional. First, it may result in fresh perspectives or, at its finest, the identification of the physics' fundamental rules. In the latter, it may result in the inclusion of elements that are irrelevant, which is generally discouraged because they do not significantly affect the prediction made by the model as a whole. Thus, choosing properties must first be guided by physical principles that have been demonstrated to apply specifically to the structure-property correlations of the materials being considered. The design and discovery of data-driven TE materials continue despite significant advancements in the development of thermoelectric materials databases due to the lack of diverse datasets containing essential material descriptors, materials synthesis parameters and sufficiently large experimental data volumes.

### *3.3.3 Pearson correlation*

The linear relationship between two random variables is denoted by the Pearson correlation or Pearson correlation coefficient. It is defined as follows [64]

$$\rho_{X,Y} = \frac{\mathrm{cov}(X, Y)}{\sigma_X \sigma_Y} \tag{7}$$

where cov represents the covariance of random variables X and Y and $\sigma_X$, $\sigma_?$ represents the standard deviation of random variables. In the field of machine learning, the Pearson correlation is frequently used for feature selection. A positive value close to 1 indicates that two variables are closely linearly connected, and one can be left out of a machine learning model to save computing costs. Furthermore, the ability of the descriptors to predict the goal linearity may be demonstrated by obtaining the Pearson correlation coefficient between material descriptors and desired qualities. A significant Pearson correlation coefficient was found between the power factor and both the electronic bandgap and carrier effective mass in Ref. [64], which used linear regression to determine the material power factor.

#### *3.3.4 Auto encoders*

Autoencoders are frequently used to transform the initial feature vectors into a lower dimensional vector known as a hidden vector, serving the same goal as PCA as an unsupervised learning technique. In order to recover high-level representations of the original characteristics, autoencoders train neural networks with identically sized input and output layers. To reduce the inaccuracy of the network output to its input, an autoencoder is trained. The benefit of autoencoders is their capacity for nonlinear transformation of the original feature vectors.

## 4. Literature review: machine learning techniques on thermoelectric materials

ML-based technologies are becoming more and more crucial in the field of TE materials due to the abundance of data from high-throughput investigations. For example, researchers use machine learning to uncover new TE materials [65] and estimate TE parameters such as band gap [66, 67], thermal conductivity [68], and Seebeck coefficient [69]. In the field of TE materials, most of the works mainly focus on improving the accuracy of predictive models [65]. High-frequency nonlinear models have been shown to outperform linear models [70–72]. However, most nonlinear machine learning algorithms are often treated as black boxes as they are too complex and inexplicable to humans, hindering the widespread adoption of machine learning. In this section, we would like to present a comprehensive overview of the research in the thermoelectric field of machine learning, together with referencing the important works previously reported in the literature. In recent years, ML has spurred widespread application in the field of materials and chemical sciences, attributed to the rapid development of artificial intelligence technology, especially machine learning methods, its high efficiency and informativeness [21, 22, 73, 74]. Materials using ML have been reported for numerous studies on thermoelectricity.

### 4.1 Machine learning studies focus on electrical transport properties

In the literature adopting high-throughput first-principle calculations offers the largest computational database of handling properties of approximately 48,000 materials [27, 54, 75–79]. In these studies, the band structure of materials is computed using Boltzmann transport theory to determine TE-related parameters such as electronic conductivity, electronic thermal conductivity and Seebeck coefficient. Also, most of the data is saved on the Material Project website, as given in **Figure 8,** including its database entries.

This database also covers the transport properties of materials at various constant doping carrier concentrations, Fermi energies and temperatures. It has been proven by further research that these calculation results have a fair agreement with the experimentally measured maximum Seebeck coefficient. This reliable and abundant

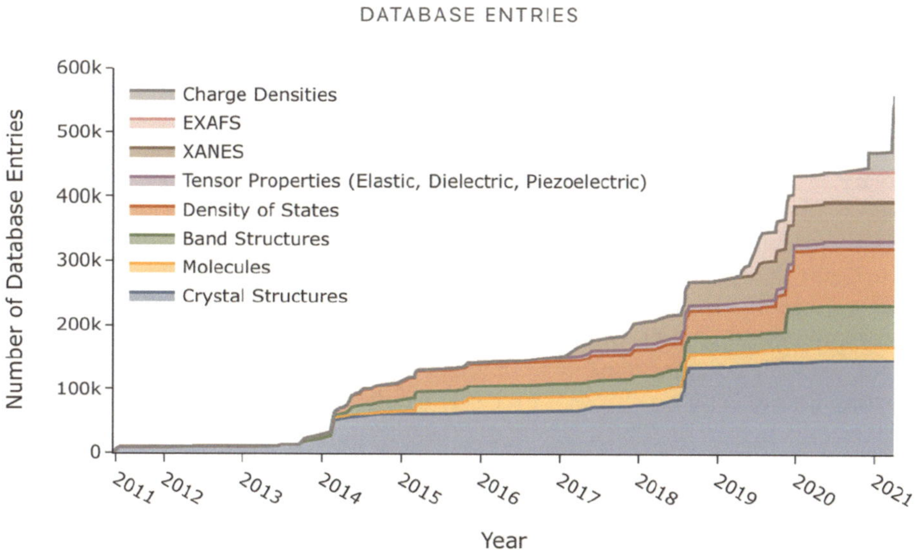


**Figure 8.**
*A snapshot of the Materials Project website and summary of its database contents.*

database is a valuable resource for machine learning-based TE material exploration techniques [80, 81]. TE features obtained from both experimental characterizations and/or theoretical calculations are crucial data sources for the machine learning process to discover new efficient TE materials. The developed machine learning model is a powerful tool for stoichiometry and nanostructure optimization for TE materials.

In the literature, machine learning techniques are used in the calculation and/or estimation of thermoelectric power factor that includes two crucial electrical transport properties: the Seebeck coefficient and electrical conductivity. Some popular ways for improving PF are band engineering [9, 82], modulation doping [6, 66, 67] and altering the effective mass of the energy band [83]. Doping is a well-known method for improving material TE characteristics, and following this method could lead to the discovery of new and efficient TE materials [84–87]. We can list a few important studies from the literature as follows; Wang et al. adopted a machine-learning technique to optimize the Cu content in Cu-doped $Bi_2Te_{2.85}Se_{0.15}$ [88]. The experimentally measured ZT with variable Cu content was used as a label for an ANN. The resulting model with a correlation coefficient of 0.99 shows excellent accuracy. Also, Hou et al. used the machine learning-based framework to discover the suitable Al/Si ratio in $Al_2Fe_3Si_3$ for TE applications [89]. To develop the model, the experimentally determined power factor for the unsynthesized materials predicted by the machine learning model was used, and the optimum material ratio increases the power factor by approximately 40%.

The ideal internal stress for TE materials was also determined using machine learning techniques. The link between XRD (X-ray diffraction) and the Seebeck coefficient of materials was discovered by Saaki et al. using machine learning [90]. Ideal stresses of 3–4% and 1–2% along the a and c axes, respectively, are predicted by the trained model to significantly increase the Seebeck coefficient.

The thermal conductivity of a material based on chemical elements is an important property for the Seebeck coefficient estimation at all temperatures, according to Furmanchuk et al.'s proposed ML solution, which can predict the Seebeck coefficients of crystalline materials in the temperature range of 300–1000 K [69]. In addition to the importance of the attribute, certain ML models may explicitly offer equations for compound attributes and identifiers. The discovered approach may be used to discern between positive and negative correlations between all descriptors and targets by examining the coefficients of the formulae.

Power factor, band gap and charge carrier effective mass have been shown to positively correlate by Wang et al. using high-throughput *ab initio* calculations and regression analysis [23]. They discovered that atoms per unit cell with many different materials typically had a high power factor.

The Seebeck coefficient, electrical conductivity, thermal conductivity, and band gap are used to determine the TE potential of a material in a web-based recommendation engine developed by Oliynyk et al. [19]. With no structural input on more than 400,000 possible combinations of elements, our Heusler discovery engine surpasses competing methods by quickly and precisely predicting the occurrence of Heusler vs. non-Heusler compounds. The model has a 0.94 actual positive rate.

Due to its great precision and speed, applications of ML in thermoelectric materials are being researched more and more. By producing attributes from the chemical formula that was proven by experiment, Iwasaki et al. published the ML model, which sped up the discovery of new candidate materials [91]. Descriptors for training the ML model were automatically produced from the composition using a composition-based feature vector (CBFV) in yet another study for the spin-driven thermoelectric effect device [92]. The findings demonstrated the significance of certain parameters for thermopower, including atomic weight, spin and orbital angular momentum. Wang et al. also used ML to study the $Cu_xBi_2Te_{2.85+y}Se_{0.15}$ system [53]. Principal component analysis (PCA) and a regression technique were used to study the relationship between microstructure and thermoelectric qualities. It was also shown that ML can build experimental setups to obtain a high ZT value in addition to forecasting the features of novel materials.

An effective method for determining the $Al_2Fe_3Si_3$ thermoelectric compound's ideal chemical composition was described by Hou et al. [20]. The Bayesian Optimization (BO) algorithm allows for successful application of machine learning to the experiment. When compared to the sample with an initial Al/Si ratio of 0.9, the power factor may be increased by roughly 40%. The framework of this study, according to the authors, might also be used for $Al_2Fe_3Si_3$ that has been exogenously doped.

The most typical method for enhancing ZT is to exogenously introduce certain elements to the BiCuSeO structure in order to lower thermal conductivity, raise carrier concentration and enhance electrical transport characteristics. With so many chemicals on the market, however, painstaking testing is required. As a result, using ML to direct the effective doping of BiCuSeO may be a smart approach to finding a solution [93–95].

Iwasaki et al. used supervised ML models to establish key physical parameters controlling the spin-driven thermoelectric effect and proposed a new material that shows promising results [44]. They established the fundamental physical parameters governing spin-driven thermoelectric (STE) materials using machine learning modeling. Their real material synthesis, which was guided by the models, resulted in the discovery of a novel STE material with a thermopower order of magnitude greater than that of the current generation of STE devices.

In 2016, Fan et al. proposed a mathematical model to calculate the optimal length and cross-sectional area of the thermoelectric generator (TEG) to maximize the power output. They found that the maximum power was obtained from the TEG at the optimum length-to-sectional area ratio under convective thermal boundary conditions [96].

In another TEG study done by Wu et al., they used a local optimization method to maximize the efficiency of a segmented TEG by adjusting the thermoelement cross-sectional area and the thickness of the segment; thereby, the total yield from TEG was 23.72% [97]. In the work of Ferreira-Teixeira and Pereira's thermocouples (TCs), made up of a p and an n-type leg, and thermoelectric devices with various geometries are numerically modeled using the COMSOL Multiphysics programme to find an optimized geometry. They reported that the optimal ratio between thermoelectric height and width should be $5 \times 10^{-3}$ [98]. They also stated that the optimal height ratio between the Cu contacts and the thermoelectric foot height was 40. The impacts of structural factors and thermodynamic boundary conditions on the output performance were examined in the pattern of identical p-n segmentation ratios by Ma et al. [99]. They reported that longer thermoelectric elements and greater heat transfer coefficients increase the ideal percentage of medium-temperature material ($CoSb_3$), whereas the cross-section area has no effect. The second pattern then examined the power improvement capability in light of the differences in the properties of p-type and n-type materials. Comparing the maximum output power to the segmented model's initial value, there has been an improvement of about 13.8%. Last but not least, the use of the best-segmented ratio design in a thermoelectric generator system showed improved performance and boosted output power by 6.8%. Kim et al. estimated the performance of a TEG running on a diesel engine using ANNs implemented using Python code [100]. Validation studies found a 3.49% difference between the output power of the experimental and predicted TEG. Wang et al. presented a fast and accurate DL model to predict the performance of TEGs [101]. The proposed deep learning model improved the power output of TEG by 182%. Kishore et al. provided ANN models that can predict TEG performance and found that two hidden layer ANNs with six neurons in each layer were most efficient in predicting the performance of TEG [102]. The optimum ANN model estimated the power and efficiency of the TEG with an accuracy of $\pm 0.1$ W and $\pm 0.2\%$, respectively, in under 26.4 microseconds per data point compared to the 6 minutes required by traditional finite element simulations. Input parameters are leg length, leg cross-sectional area and external resistance. They also explained that increasing the number of neurons per layer above gold does not improve the prediction accuracy of the ANN.

In the study by Zhu et al., a DL technique is used to forward simulate the maximum power output and efficiency of a thermoelectric generator as well as its use in generator design and optimization [103] after being trained on a dataset made up of 5000 3-D finite element method-based simulations, artificial neural networks with five layers and 400 neurons per layer displayed extraordinarily high prediction accuracy of over 98%. Furthermore, they might function under situations of continuous heat flux and temperature difference while taking into consideration thermoelectric phenomena such as contact electrical resistance and surface heat transfer.

Ang et al. predicted a TEG's energy output in its operational environment using an ANN model [104]. A multilayer perceptron (MLP) was trained in a supervised manner and evaluated on the dataset created using a verified finite volume approach to forecast the energy generated. Their model could also conduct reverse ANN to predict the input value when given an output value, in addition to forecasting the output values.

### 4.2 Machine learning studies focus on thermal transport properties

To make TE energy an economically feasible alternative for waste heat recovery, TE materials with $ZT > 1$ are necessary. The construction of phonon glass-electron crystal structures that enable the separation of electron and phonon transport characteristics has been the main experimental emphasis in the investigation of oxide TE materials [105]. The development of thermoelectric oxides has mostly been focused on techniques that enhance the hierarchical scattering of phonons. The most popular method for starting hierarchical phonon scattering has been the use of sintering additives [106–108]. The literature study makes it abundantly evident that the class of materials found for TE applications has so far been relatively constrained and that our knowledge of the electronic and phonon transport of crystalline alloys is quite constrained [109]. On the other hand, quick advancements in materials informatics have aided researchers in finding novel, promising classes of materials and establishing links between design factors and thermoelectric characteristics [65]. Large lattice parameters, a wide band gap, and a high effective hole mass are essential characteristics for nanoparticle semi-Heusler compounds to have a high TE efficiency, according to high-throughput material modeling and ML approaches. For more experimental research, new semiconductors with extremely low $\kappa_{ph}$ values have been suggested [24]. In essence, there have been a lot of theoretical and empirical attempts over the past few decades to evaluate the $\kappa_{ph}$ of various systems. A mapping between the input properties (such as the atomic mass, phonon frequency, and unit cell volume) and the target property $\kappa_{ph}$ may be produced using ML using this data. In contrast to first-principles calculations and MD simulations, the data-driven ML models enable high-throughput evaluation of $\kappa_{ph}$, which has outstanding predictive potential for systems inside and outside the training set.

Juneja et al. performed high-throughput *ab initio* calculations on 195 binary, ternary, and quaternary molecules in the dataset [110]. Calculations were made to determine the lattice thermal conductivity $\kappa_{ph}$ values, which range over three orders of magnitude, for 120 dynamically stable non-metallic compounds. 11 ultrahigh and 15 ultralow $\kappa_{ph}$ materials are found among them. According to an investigation of the created property map for this dataset, $\kappa_{ph}$ strongly depends on four basic descriptors: average atomic mass, maximum phonon frequency, integrated Grüneisen parameter up to 3 THz, and unit cell volume. An ML model based on Gaussian process regression was created using these descriptors. The model's exceptionally low root mean square error of 0.21 predicted log-scaled $\kappa_{ph}$.

Zhang and Ling suggested including a rough estimate of the target feature using low-quality models as a way to improve the accuracy of ML models applied to small datasets. By including the empirical abundance model values of $\kappa_{ph}$ as descriptors in the ML model, they achieved high accuracy in estimating $\kappa_{ph}$ [111]. The link between the degree of freedom (DoF) of the model and the accuracy of prediction was revealed by their investigation as a significant occurrence when the model is trained using limited accessible materials data. The emergence of the precision-DoF relationship, which resulted from the statistical bias-variance tradeoff, limits the accuracy of prediction in unknowable domains. They also suggested employing the crude estimating of property in the feature space as a way to increase accuracy without increasing DoF. The incorporation of crude estimate significantly increased the predicted accuracy of ML models in three case studies, illuminating the applicability of the suggested method for building precise ML models from sparse materials data.

Chen et al. developed an ML-based model and employed sophisticated general property engineering technology in conjunction with the Gaussian process regression technique to estimate the phonon contribution value of inorganic materials [112]. Using a benchmark data set of around 100 inorganic materials that have been experimentally characterized, they developed an ML-based model to quickly and correctly discover inorganic materials. Along with the Gaussian process regression approach, they applied sophisticated and ubiquitous feature engineering techniques.

Juneja et al. combined ML with high-throughput computation to create regression models to predict $\kappa_{ph}$ of inorganic compounds. They also used the maximum phonon frequency and the integrated Grüneisen parameter as descriptors to construct ML models to estimate $\kappa_{ph}$ [113, 114]. Both ML models used complex derived features as descriptors in their ML models to predict $\kappa_{ph}$, which limited their use in the early stages of material selection and design. ML models based on characteristic material properties should be used effectively in the discovery of new materials and shorten the design cycle time.

Jaafreh et al. performed high-throughput *ab initio* calculations on 195 binary, ternary, and quaternary molecules in the dataset [34]. Calculations are made to determine the lattice thermal conductivity $\kappa_{ph}$ values, which range over three orders of magnitude, for 120 dynamically stable non-metallic compounds. 11 ultrahigh and 15 ultralow l materials are found among them. According to an investigation of the created property map for this dataset, $\kappa_{ph}$ strongly depends on four basic descriptors: average atomic mass, maximum phonon frequency, integrated Grüneisen parameter up to 3 THz, and unit cell volume. An ML model based on Gaussian process regression was created using these descriptors. The model's exceptionally low root mean square error of 0.21 predicts log-scaled $\kappa_{ph}$.

The optimization of random multilayer structures (RMLs) is vital for achieving ultralow thermal conductivity, which is critical for a wide range of applications including thermoelectric materials. Chakraborty et al. found some critical criteria for assessing disorder in RML layer thicknesses [115]. Classical molecular dynamics simulations of hypothetical Lennard-Jones RMLs supported our ability to associate these disorder characteristics with thermal conductivity. Furthermore, they demonstrated that these metrics may be used as features in physics-based machine-learning models to predict the lattice thermal conductivity of RMLs with greater accuracy and efficiency.

Half-Heusler compounds were utilized as prototype examples by Liu et al. to show how a compressed-sensing approach may be applied to quickly and accurately assess lattice thermal conductivity, as realized by a physically interpretable descriptor [116]. Seventy-five half- and 15 full-Heusler compounds' thermal conductivities were predicted using the descriptor, and the results show good agreement with explicit first-principles findings. The descriptor was further improved by supplying only the fundamental characteristics of the constituent atoms, which helped hasten the search for materials with the proper thermal conductivity.

The heat conductivity of two-dimensional materials such as graphene may be easily controlled by inserting holes, the density and distribution of which are crucial characteristics. To investigate the link between hole distribution and thermal conductivity decrease in monolayer graphene, Wan et al. used an inverse design process based on machine learning [117]. According to their method, the best distribution for reducing thermal conductivity in porous graphene is one in which holes are randomly distributed transverse to the direction of heat flow yet exhibit some periodicity along the direction of heat flow.

Carbon honeycombs (CHCs) and boron nitride honeycombs (BNHCs) have been revealed to have identical molecular architectures but distinct thermal characteristics. Thus, hybrid carbon-boron nitride honeycombs (C-BNHCs) with adjustable thermal conductivity may be created by correctly patching together CHCs and BNHCs. Du et al. used the ML approach in conjunction with molecular dynamics simulations to examine the thermal transport property of C-BNHCs, as well as to design C-BNHC structures for specified thermal conductivity [118]. In the inverse design of C-BNHCs with any given thermal conductivity, their ML-based technique demonstrated remarkable accuracy and efficiency.

Zhu et al. estimated the thermal conductivity of all known inorganic materials in the Inorganic Crystal Structure Database using a combination of graph neural networks and random forest techniques, then charted the structural chemistry into extended van-Arkel triangles [119]. Using the newly constructed map and their theoretical tool, we identify rare-earth chalcogenides as promising possibilities, with ZT values more than 1.0.

Ju et al. demonstrated that when lower-order feature qualities present in big data are appropriately selected and applied to transfer learning, large data may supplement small data for accurate predictions [120]. A neural network was used to directly connect the crystal information and thermal conductivity by transferring descriptors obtained from a pre-trained model for the feature property. Successful transfer learning demonstrated extrapolative prediction abilities and revealed descriptors for lattice anharmonicity. The resultant model was used to screen over 60,000 chemicals for unique crystals that might be used as diamond substitutes.

In order to detect unexpected lattice thermal conductivity $\kappa_{ph}$ enhancement in aperiodic superlattices versus periodic superlattices, Chowdhury and Ruan demonstrated a general-purpose adaptive ML-accelerated search process [121]. This process has implications for the thermal management of multilayer-based electronic devices. They employed molecular dynamics simulations to calculate $\kappa_{ph}$ with great precision, as well as a convolutional neural network (CNN) to forecast $\kappa_{ph}$ for a large number of structures. They repeatedly discovered aperiodic superlattices (SLs) with structural properties leading to locally improved heat transport and used them as extra training data for the CNN to enable accurate prediction for the target unknown SLs. Because of the existence of closely spaced surfaces, the detected structures displayed higher coherent phonon transport.

## 5. Summary and future perspectives

Thermoelectric materials are particularly beneficial in a variety of applications due to their non-toxic, low-cost, earth-plentiful, low-density and ecologically acceptable properties. To address today's energy issues, research on efficient thermoelectric materials is becoming more and more important. The development of highly efficient thermoelectric materials has advanced significantly over the past few decades in both theoretical and practical investigations. Recent developments in nanotechnology, in particular, have introduced approaches that hold promise for improving the thermoelectric efficiency of basic systems. Although tremendous progress has been achieved in the literature, the scientific community is still concentrating its efforts on the discovery of a new generation of thermoelectric materials with the highest efficiency and applicability to everyday life. Therefore, the most important current goal in the thermoelectric research area is to find new and innovative thermoelectric material systems.

The purpose of this chapter was to discuss current advances in machine learning-assisted thermoelectric material discovery. Training from the correlation between thermoelectricity and material transport properties, machine learning might provide an advantageous thermoelectric material discovery tool for new chemical composition, nano-structural design, stoichiometry optimization and other applications. This newly acquired data might be used to extend the thermoelectric database and increase the training performance of the machine learning model. For further investigation, the active learning technique is advised. The given working framework may be relied on when using AI-guided data-driven methodologies for thermoelectric material discovery. Since the majority of pertinent studies focused only on using machine learning technologies to discover and create materials with excellent thermoelectricity, it is suggested that additional benefits, such as non-toxicity and earth-abundance, be taken into account as the additional output of the thermoelectric discovery tool.

## Acknowledgements

Ö.C. Yelgel acknowledges the support from the University of Manchester, the National Graphene Institute and the School of Physics and Astronomy.

## Conflict of interest

The authors declare no conflict of interest.

## Author details

Ebrar Yildirim and Övgü Ceyda Yelgel*
Department of Electrical-Electronics Engineering, Recep Tayyip Erdoğan University, Rize, Turkey

*Address all correspondence to: oceyda.yelgel@erdogan.edu.tr

## References

[1] Rowe DM. Thermoelectrics Handbook. Boca Raton: CRC Press; 2005

[2] Stabler FR. Automotive applications for high efficiency thermoelectrics. In: High Efficieny Workshop. San Diego, CA. 2002. p. 24

[3] Zhao LD, Wu HJ, Hao SQ, Wu CI, Zhou XY, Biswas K, et al. All-scale hierarchical thermoelectrics: MgTe in PbTe facilitates valence band convergence and suppresses bipolar thermal transport for high performance. Energy Environ Science. 2013;**6**:3346

[4] Yelgel ÖC. Theoretical study of thermoelectric properties of p-type $Mg_2Si_{1-x}Sn_x$ solid solutions doped with Ga. Journal of Alloys and Compounds. 2017;**691**:151

[5] Xie W, Weidenkaff A, Tang X, Zhang Q, Poon J, Tritt TM. Recent advances in nanostructured thermoelectric half-Heusler compounds. Nanomaterials. 2012;**2**(4):379

[6] Liu W, Tan X, Yin K, Liu H, Tang X, Shi J, et al. Convergence of conduction bands as a means of enhancing thermoelectric performance of n-type $Mg_2Si_{1-x}Sn_x$ solid solutions. Physical Review Letters. 2012;**108**:166601

[7] Gan Y, Wang G, Zhou J, Sun Z. Prediction of thermoelectrics performance for layered IV-V-VI semiconductors by high-throughput ab-inito calculations and machine learning. NPJ Computational Materials. 2021;7:176

[8] Xi L, Pan S, Li X, Xu Y, Ni J, Sun X, et al. Discovery of high-performance thermoelectric chalcogenides through reliable high-throughput material screening. Journal of the American Chemical Society. 2018;**140**: 10785-10793

[9] Graziosi P, Kumarasinghe C, Neophytou N. Impact of the scattering physics on the power factor of complex thermoelectric materials. Journal of Applied Physics. 2019;**126**: 155701

[10] Lu N, Han G, Feng Y, Sun Y, Lin G. Artificial intelligence assisted thermoelectric materials and discovery. Research Square. 2022;**1**:1-13

[11] Balachandran PV, Xue D, Theiler J, Hogden J, Lookman T. Adaptive strategies for materials design using uncertainties. Scientific Reports. 2016;**6**: 19660

[12] Bassman L, Rajak P, Kalia RK, Nakano A, Sha F, Sun J, et al. Active learning for accelerated design of layered materials. NPJ Computational Materials. 2018;**4**:74

[13] Sheng Y, Wu Y, Yang J, Lu W, Villars P, Zhang W. Active learning for the power factor prediction in diamond-like thermoelectric materials. NPJ Computational Materials. 2020;**6**:171

[14] De Witte J. Data-efficient discovery of thermoelectric materials using deep learning, Ghent University Faculty of Engineering and Architecture Master's Thesis. 2020

[15] Han G, Sun Y, Feng Y, Lin G, Lu N. Artificial intelligence guided thermoelectric materials design and discovery. Advance Electronic Materials. 2023;**9**:2300042

[16] Han G, Sun Y, Feng Y, Lin G, Lu N. Machine learning regression guided

thermoelectric materials discovery. ES Materials and Manufacturing. 2021;**14**: 20-35

[17] Xu Y, Xiangmeng W, Li X, Xi L, Ni J, Zhu W, et al. New materials band gap prediction based on the high-throughput calculation and the machine learning. Scientia Sinica Technologica. 2019;**49**: 44-54

[18] Wang X, Xu Y, Yang J, Ni J, Zhang W, Zhu W. ThermoEPred-EL: Robust bandgap predictions of chalcogenides with diamond-like structure via feature cross-based stacked ensemble learning. Computational Materials Science. 2019;**169**:109117

[19] Oliynyk AO, Antono E, Sparks TD, Ghadbeigi L, Gaultois MW, Meredig B, et al. High-throughput machine learning-driven synthesis of full-heusler compounds. Chemistry of Materials. 2016;**28**(20):7324-7331

[20] Hou Z, Takagiwa Y, Shinohara Y, Xu Y, Tsuda K. Machine-learning-assisted development and theoretical consideration for the $Al_2Fe_3Si_3$ thermoelectric material. ACS Applied Materials & Interfaces. 2019; **11**(12):11545-11554

[21] Le T, Epa VC, Burden FR, Winkler DA. Quantitative structure-property relationship modelling of diverse materials properties. Chemical Reviews. 2012;**112**(5):2889-2919

[22] Pilania G, Wang C, Jiang X, Rajasekaran S, Ramprasad R. Accelerating materials property predictions using machine learning. Scientific Reports. 2013;**3**(1):1-6

[23] Wang S, Wang Z, Setyawan W, Mingo N, Curtarolo S. Assessing the thermoelectric properties of sintered compounds via high-throughput ab-initio calculations. Physical Review X. 2011;**1**(2):021012

[24] Carrete J, Li W, Mingo N, Wang S, Curtarolo S. Finding unprecedentedly low- thermal-conductivity half-heusler semiconductors via high-throughput materials modelling. Physical Review X. 2014;**4**(1):011019

[25] Gorai P, Gao D, Ortiz B, Miller S, Barnett SA, Mason T, et al. TE Design Lab: A virtual laboratory for thermoelectric material design. Computational Materials Science. 2016; **112**:368-376

[26] TE Design Lab. Available from: https://tedesignlab.org [Accessed: March 13, 2022]

[27] Ricci F, Chen W, Aydemir U, Snyder GJ, Rignanese GM, Jain A, et al. An ab initio electronic transport database for inorganic materials. Scientific Data. 2017;**4**:170085

[28] Data from: An Ab initio electronic transport database for inorganic materials. Available from: https://datadryad.org/stash/dataset/doi:10.5061/dryad.gn001 [Accessed: April 02, 2023]

[29] Chen L, Tran H, Batra R, Kim C, Ramprasad R. Machine learning models for the lattice thermal conductivity prediction of inorganic materials. Computerised Material Science. 2019; **170**:109155

[30] Katsura Y, Kumagai M, Kodani T, Kaneshige M, Ando Y, Gunji S, et al. Data-driven analysis of electron relaxation times in PbTe type thermoelectric materials. Science and Technology of Advanced Materials. 2019;**20**:511-520

[31] Starrydata Dataset. Available from: https://github.com/starrydata/starrydata_datasets [Accessed: March 13, 2022]

[32] Priya P, Aluru N. Accelerated design and discovery of perovskites with high

conductivity for energy applications through machine learning. NPJ Computational Materials. 2021;7:90

[33] Data from: Accelerated design and discovery of perovskites with high conductivity for energy applications through machine learning. Available from: https://figshare.com/s/10b18051e26fa4d4f18c [Accessed: April 02, 2023]

[34] Jaafreh R, Kang YS, Hamad K. Lattice thermal conductivity: An accelerated discovery guided by machine learning. ACS Applied Materials & Interfaces. 2021;**13**:57204-57213

[35] Miyazaki H, Tamura T, Mikami M, Watanabe K, Ide N, Ozkendir OM, et al. Machine learning based prediction of lattice thermal conductivity for half-Heusler compounds using atomic information. Scientific Reports. 2021;**11**:1-8

[36] Yao M, Wang Y, Li X, Sheng Y, Huo H, Xi L, et al. Materials informatics platform with three dimensional structures, workflow and thermoelectric applications. Scientific Data. 2021;**8**:236

[37] MatHub-3d. Available from: http://www.mathub3d.net/materials/matdb [Accessed: April 02, 2023]

[38] Tranas R, Lvvik OM, Tomic O, Berland K. Lattice thermal conductivity of half- Heuslers with density functional theory and machine learning: Enhancing predictivity by active sampling with principal component analysis. Computational Materials Science. 2022;**202**:110938

[39] Thursby JG, Schmidt P. Some properties of tests for specification error in a linear regression model. Journal of the American Statistical Association. 1977;**72**(359):635-641

[40] Roekeghem A, Carrete JU, Oses C, Curtarolo S, Mingo N. High-throughput computation of thermal conductivity of high-temperature solid phases: The case of oxide and fluoride perovskites. Physical Review X. 2016;**6**:041061

[41] De Jong M, Chen W, Notestine R, Persson K, Ceder G, Jain A, et al. A statistical learning framework for materials science: Application to elastic moduli of k-nary inorganic polycrystalline compounds. Scientific Reports. 2016;**6**:34256

[42] Miller SA, Dylla M, Anand S, Gordiz K, Snyder GJ, Toberer ES. Empirical modeling of dopability in diamond-like semiconductors. NPJ Computational Materials. 2018;**4**:1-8

[43] Li Z, Xu Q, Sun Q, Hou Z, Yin WJ. Stability engineering of halide perovskite via machine learning. Advanced Functional Materials. 2019;**29**:1807280

[44] Iwasaki Y, Takeuchi I, Stanev V, Kusne AG, et al. Machine-learning guided discovery of a new thermoelectric material. Scientific Reports. 2019;**9**:1-7

[45] Tian Y, Shi Y, Liu X. Recent advances on support vector machines research. Technological and Economic Development of Economy. 2012;**18**(1):5-33

[46] Ray S. A quick review of machine learning algorithms. In: International Conference on Machine Learning. Faridabad, India: Big Data Cloud Parallel Computing Communication; 2019. pp. 35-39

[47] Allers J, Harvey J, Garzon F, Alam T. Machine learning prediction of self diffusion in Lennard-Jones fluids. Journal of Chemical Physics. 2020;**153**:034102

[48] Breiman L. Random forests. Machine Learning. 2001;**45**:5-32

[49] Friedman JH. Greedy function approximation: A gradient boosting machine. The Annals of Statistics. 2001; **29**(5):1189-1232

[50] Gaultois MW, Oliynyk AO, Mar A, Sparks TD, Mulholland GJ, Meredig B. Web-based machine learning models for real-time screening of thermoelectric materials properties. APL Materials. 2016;**4**:053213

[51] Furmanchuk A, Saal J, Doak JW, Olson GB, Choudhary A, Agrawal A. Prediction of seebeck coefficient for compounds without restriction to fixed stoichiometry: A machine learning approach. Journal of Computational Chemistry. 2018;**39**:191-201. DOI: 10.1002/jcc.25067

[52] Wu YJ, Fang L, Xu Y. Predicting interfacial thermal resistance by machine learning. NPJ Computational Materials. 2019;**5**:56

[53] Wang ZL, Adachi Y, Chen ZC. Processing optimization and property predictions of hot-extruded Bi–Te–Se thermoelectric materials via machine learning. Advanced Theory and Simulations. 2020;**3**:1900197

[54] Han G, Sun Y, Feng Y, Lin G, Lu N. Machine learning regression guided thermoelectric materials discovery – A review. ES Materials Manufacturing. 2021;**14**:20-35

[55] Hornik K, Stinchcombe M, White H. Multilayer feedforward networks are universal approximators. Neural Networks. 1989;**2**:359-366

[56] LeCun Y, Bengio Y, Hinton G. Deep learning. Nature. 2015;**521**: 436-444

[57] Wang Y, Jiang Y, Lan J. FCNN: An efficient intrusion detection method based on raw network traffic. Security and Communication Networks. 2021;**13**: 5533269

[58] Hochreiter S, Schmidhuber J. Long short-term memory. Neural Computation. 1997;**9**(8):1735-1780

[59] Shrestha A, Mahmood A. Review of deep learning algorithms and architectures. IEEE Access. 2019;**7**: 53040-53065. DOI: 10.1109/ ACCESS.2019.2912200

[60] Shlens J. A tutorial on principal component analysis. arXiv. 2014;**1404**: 1100

[61] Wagner N, Rondinelli JM. Theory-guided machine learning in materials. Frontiers in Materials. 2016;**3**:28

[62] Natekin A, Knoll A. Gradient boosting machines, a tutorial. Frontiers in Neurorobotics. 2013;**7**:21

[63] Ong SP, Richards WD, Jain A, Hautier G, Kocher M, Cholia S, et al. Python materials genomics (pymatgen): A robust, open-source python library for materials analysis. Computational Materials Science. 2013;**68**:314-319

[64] Mbaye MT, Pradhan SK, Bahoura M. Data-driven thermoelectric modelling: Current challenges and prospects. Journal of Applied Physics. 2021;**130**: 190902

[65] Wang T, Zhang C, Snoussi H, Zhang G. Machine learning approaches for thermoelectric materials research. Advanced Functional Materials. 2020; **30**(5):1906041

[66] Zhang J, Liu R, Cheng N, Zhang Y, Yang J, Uher C, et al. High-performance

pseudocubic thermoelectric materials from non cubic chalcopyrite compounds. Advanced Materials. 2014; **26**:3848-3853

[67] Pei Y, Shi X, LaLonde A, Wang H, Chen L, Snyder GJ. Convergence of electronic bands for high performance bulk thermoelectrics. Nature. 2011;**473**: 66-69

[68] Wang X, Zeng S, Wang Z, Ni J. Identification of crystalline materials with ultra-low thermal conductivity based on machine learning study. The Journal of Physical Chemistry C. 2020; **124**:8488-8495

[69] Furmanchuk A, Saal JE, Doak JW, Olson GB, Choudhary A, Agrawal A. Prediction of Seebeck coefficient for compounds without restriction to fixed stoichiometry: A machine learning approach. Journal of Computational Chemistry. 2018;**39**:191-202

[70] Gladkikh V, Kim DY, Hajibabaei A, Jana A, Myung CW, Kim KS. Machine learning for predicting the band gaps of $ABX_3$ perovskites from elemental properties. The Journal of Physical Chemistry C. 2020;**124**:8905-8918

[71] Ramprasad R, Batra R, Pilania G, Mannodi-Kanakkithodi A, Kim C. Machine learning in materials informatics: Recent applications and prospects. npj Computational Materials. 2017;**3**:54

[72] Schmidt J, Marques MRG, Botti S, Marques MAL. Recent advances and applications of machine learning in solid-state materials science. NPJ Computational Materials. 2019;**5**:83

[73] Panapitiya G, Avendano FG, Ren P, Wen X, Li Y, Lewis JP. Machine learning prediction of CO adsorption in thiolated, Ag-alloyed Au nanoclusters. Journal of the American Chemical Society. 2018; **140**(50):17508-17514

[74] Rajan AC, Mishra A, Satsangi S, Vaish R, Mizuseki H, Lee KR, et al. Machine-learning-assisted accurate band gap predictions of functionalized MXene. Chemistry of Materials. 2018; **30**(12):4031-4038

[75] Hao Q, Xu D, Lu N, Zhao H. High-throughput ZT predictions of nanoporous bulk materials as next-generation thermoelectric materials: A material genome approach. Physical Review B. 2016;**93**:205206

[76] Curtarolo S, Hart GLW, Nardelli MB, Mingo N, Sanvito S, Levy O. The high-throughput highway to computational materials design. Nature Materials. 2013;**12**:191-201

[77] Greeley J, Jaramillo TF, Bonde J, Chorkendorff I, Jens K. Computational high-throughput screening of electrocatalytic materials for hydrogen evolution. Nature Materials. 2006;**5**: 909-913

[78] Bhattacharya S, Chmielowski R, Dennler G, Madsen GKH. Novel ternary sulfide thermoelectric materials from high throughput transport and defect calculations. Journal of Materials Chemistry A. 2016;**4**:11086-11093

[79] Liu Z, Fu B, Yi X, Yuan G, Wang J, Ferguson I. Co-doping of magnesium with indium in nitrides: First principle calculation and experiment. RSC Advances. 2016;**6**:5111-5115

[80] Bishara D, Xie Y, Liu WK, Li S. A state-of-the-art review on machine learning-based multiscale modelling, simulation, homogenization and design of materials. Archives of Computational Methods in Engineering. 2023;**30**: 191-222

[81] Choudhary K, Garrity KF, Tavazza F. Data-driven discovery of 3D and 2D thermoelectric materials. Journal of Physics: Condensed Matter. 2020;**32**: 475501

[82] Graziosi P, Kumarasinghe C, Neophytou N. Material descriptors for the discovery of efficient thermoelectrics. ACS Applied Energy Materials. 2020;**3**:5913-5926

[83] Fu C, Zhu T, Liu Y, Xie H, Zhao X. Band engineering of high performance p-type FeNbSb based half-Heusler thermoelectric materials for figure of merit zT > 1. Energy & Environmental Science. 2015;**8**:216e20

[84] Wang ZL, Yokoyama Y, Onda T, Adachi Y, Chen ZC. Influence of algorithm parameters of Bayesian optimization, genetic algorithm, and particle swarm optimization on their optimization performance. Advanced Theory and Simulations. 2019;**5**:1900079

[85] Wang B, Kucukgok B, He Q, Melton AG, Leach J, Udwary K, et al. Thermoelectric properties of undoped and Si-doped bulk GaN. MRS Online Proceedings Library. 2013;**1558**:903

[86] Li J, Sui J, Pei Y, Meng X, Berardan D, Dragoe N, et al. The roles of Na doping in BiCuSeO oxyselenides as a thermoelectric material. Journal of Materials Chemistry A. 2014;**2**:4903-4906

[87] Zhang D, Yang J, Jiang Q, Fu L, Xiao Y, Luo Y, et al. Improvement of thermoelectric properties of $Cu_3SbSe_4$ compound by in doping. Materials and Design. 2016;**98**:150-154

[88] Zhou B, Li S, Li W, Li J, Zhang X, Lin S, et al. Thermoelectric properties of SnS with Na-doping. ACS Applied Materials & Interfaces. 2017;**9**: 34033-34041

[89] Hou Z, Takagiwa Y, Shinohara Y, Xu Y, Tsuda K. Fe–Al–Si thermoelectric (FAST) materials and modules: Diffusion couple and machine-learning-assisted materials development. ACS Applied Materials & Interfaces. 2019;**11**: 11545-11554

[90] Sasaki M, Ju S, Xu Y, Shiomi J, Goto M. Identifying optimal strain in Bismuth telluride thermoelectric film by combinatorial gradient thermal annealing and machine learning. ACS Combinatorial Science. 2020;**22**:782-790

[91] Iwasaki Y, Takeuchi I, Stanev V, Kusne AG, Ishida M, Kirihara A, et al. Machine-learning guided discovery of a new thermoelectric material. Scientific Reports. 2019;**9**:2751

[92] Murdock RJ, Kauwe SK, Wang AYT, Sparks TD. Is domain knowledge necessary for machine learning materials properties? Integrated Materials. 2020;**9**: 221-227

[93] Li F, Ruan M, Chen Y, Wang W, Luo J, Zheng Z, et al. Enhanced thermoelectric properties of polycrystalline BiCuSeO via dual-doping in Bi sites. Inorganic Chemistry Frontiers. 2019;**6**:799-807

[94] Das S, Valiyaveettil S, Chen KH, Suwas S, Mallik R. Thermoelectric properties of Pb and Na dual doped BiCuSeO. AIP Advances. 2019;**9**:015025

[95] Feng B, Li G, Pan Z, Hu X, Liu P, Li Y, et al. Enhanced thermoelectric performances in BiCuSeO Oxyselenides via Er and 3D modulation doping. Ceramics International. 2019;**45**: 4493-4498

[96] Fan L, Zhang G, Wang R, Jiao K. A comprehensive and time-efficient model for determination of thermoelectric generator length and cross-section area.

Energy Conversion and Management. 2016;**122**:85-94

[97] Wu Y, Yang J, Chen S, Zuo L. Thermo-element geometry optimization for high thermoelectric efficiency. Energy. 2018;**147**:672-680

[98] Ferreira-Teixeira S, Pereira AM. Geometrical optimization of a thermoelectric device: Numerical simulations. Energy Conversion and Management. 2018;**169**:217-227

[99] Ma X, Shu G, Tian H, Xu W, Chen T. Performance assessment of engine exhaust-based segmented thermoelectric generators by length ratio optimization. Applied Energy. 2019;**248**:614-625

[100] Kim TY. Prediction of system-level energy harvesting characteristics of a thermoelectric generator operating in a diesel engine using artificial neural networks. Energies. 2021;**14**:2426

[101] Wang P, Wang K, Xi L, Gao R, Wang B. Fast and accurate performance prediction and optimization of thermoelectric generators with deep neural networks. Advanced Materials Technologies. 2021;**6**:2100011

[102] Kishore R, Mahajan R, Priya S. Combinatory finite element and artificial neural network model for predicting performance of thermoelectric generator. Energies. 2018;**11**:2216

[103] Zhu Y, Newbrook DW, Dai P, de Groot CHK, Huang R. Artificial neural network enabled accurate geometrical design and optimisation of thermoelectric generator. Applied Energy. 2022;**305**:117800

[104] Ang ZYA, Woo WL, Mesbahi E. Artificial neural network based prediction of energy generation from thermoelectric generator with environmental parameters. Journal of Clean Energy Technologies. 2017;**5**: 458-463

[105] He J, Liu Y, Funahashi R. Oxide thermoelectrics: The challenges, progress, and outlook. Journal of Materials Research. 2011;**26**(15): 1762-1772

[106] Wang N, He H, Ba Y, Wan C, Koumoto K. Thermoelectric properties of nb-doped srtio3 ceramics enhanced by potassium titanate nanowires addition. Journal of the Ceramic Society of Japan. 2010;**118**(1383):1098-1101

[107] Buscaglia MT, Maglia F, Anselmi-Tamburini U, Marré D, Pallecchi I, Ianculescu A, et al. Effect of nanostructure on the thermal conductivity of la-doped srtio3 ceramics. Journal of the European Ceramic Society. 2014;**34**(2):307-316

[108] Lan J, Lin YH, Liu Y, Xu S, Nan CW. High thermoelectric performance of nanostructured $In_2O_3$-based ceramics. Journal of the American Ceramic Society. 2012;**95**(8):2465-2469

[109] Minnich A, Dresselhaus MS, Ren Z, Chen G. Bulk nanostructured thermoelectric materials: Current research and future prospects. Energy & Environmental Science. 2009;**2**(5): 466-479

[110] Juneja R, Yumnam G, Satsangi S, Singh AK. Coupling the high-throughput property map to machine learning for predicting lattice thermal conductivity. Chemistry of Materials. 2009;**31**: 5145-5151

[111] Zhang Y, Ling C. A strategy to apply machine learning to small datasets in materials science. NPJ Computational Materials. 2018;**4**(1):1-8

[112] Chen L, Huan T, Batra R, Kim C, Ramprasad R. Machine learning models for the lattice thermal conductivity prediction of inorganic materials. Computational Materials Science. 2019; **170**:109155

[113] Juneja R, Singh AK. Unravelling the role of bonding chemistry in connecting electronic and thermal transport by machine learning. Journal of Materials Chemistry A. 2020a;**8**(17): 8716-8721

[114] Juneja R, Singh AK. Guided patchwork kriging to develop highly transferable thermal conductivity prediction models. Journal of Physics: Materials. 2020;**3**(2):024006

[115] Chakraborty P, Liu Y, Ma T, Guo X, Cao L, Hu R, et al. Quenching thermal transport in aperiodic superlattices: A molecular dynamics and machine learning study. ACS Applied Materials & Interfaces. 2020;**12**:8795-8804

[116] Liu J, Han S, Cao G, Zhou Z, Sheng C, Liu H. A high-throughput descriptor for prediction of lattice thermal conductivity of half-Heusler compounds. Journal of Physics D: Applied Physics. 2020;**53**:315301

[117] Wan J, Jiang JW, Park HS. Machine learning-based design of porous graphene with low thermal conductivity. Carbon. 2020;**157**:262-269

[118] Du Y, Ying P, Zhang J. Prediction and optimization of the thermal transport in hybrid carbon-boron nitride honeycombs using machine learning. Carbon. 2021;**184**:492-503

[119] Zhu Y, He R, Gong S, Xie T, Gorai P, Nielsch GJC. Charting lattice thermal conductivity for inorganic crystals and discovering rare earth chalcogenides for thermoelectrics. Energy & Environmental Science. 2021; **14**:3559-3566

[120] Ju S, Yoshida R, Liu C, Wu S, Hongo K, Tadano T, et al. Exploring diamondlike lattice thermal conductivity crystals via feature-based transfer learning. Physical Review Materials. 2021;**5**:053801

[121] Chowdhury PR, Ruan X. Unexpected thermal conductivity enhancement in aperiodic superlattices discovered using active machine learning. NPJ Computational Materials. 2022;**8**:12

Chapter 4

# Organic Thermoelectric Materials

*Sikander Azam and Muhammad Farzik Ijaz*

## Abstract

This book chapter provides a comprehensive overview of organic thermoelectric materials and their potential applications. Organic materials have recently emerged as promising candidates for thermoelectric devices due to their unique combination of electrical conductivity and thermal properties. The chapter begins by discussing the fundamental principles and mechanisms underlying the thermoelectric effect in organic materials, including the Seebeck coefficient, electrical conductivity, and thermal conductivity. It further explores various strategies employed to enhance the thermoelectric performance of organic materials, such as molecular design, doping, and nanostructuring. Additionally, the chapter highlights recent advancements in the synthesis and characterization of organic thermoelectric materials, including polymer-based systems, small organic molecules, and hybrid organic-inorganic composites. The discussion also extends to the evaluation techniques and metrics used to assess the thermoelectric efficiency of organic materials. Furthermore, the chapter sheds light on the challenges and opportunities in the field, such as stability, scalability, and cost-effectiveness, along with potential applications in energy harvesting, waste heat recovery, and wearable electronics. Overall, this book chapter aims to provide a comprehensive understanding of organic thermoelectric materials and their significant role in advancing thermoelectric technology.

**Keywords:** organic thermoelectric materials, thermoelectric effect, electrical conductivity, thermal conductivity, energy harvesting

## 1. Introduction

Thermoelectric materials have been gaining considerable attention in the field of energy conversion due to their ability to directly convert waste heat into useful electrical energy. They are materials that exhibit the thermoelectric effect, which is the generation of a voltage or a temperature difference due to a temperature gradient in the material. The efficiency of thermoelectric materials is described by the dimensionless figure of merit (ZT), which is directly proportional to the thermoelectric conversion efficiency. In recent years, organic thermoelectric materials have emerged as a promising candidate for efficient and low-cost thermoelectric energy conversion.

Organic thermoelectric materials (see **Figure 1**) are composed of organic compounds, and they have a number of advantages over traditional inorganic thermoelectric materials. First, they are lightweight, flexible, and can be easily processed into various forms, including thin films, fibers, and bulk materials. Second, they are environmentally friendly and can be produced using low-cost and sustainable

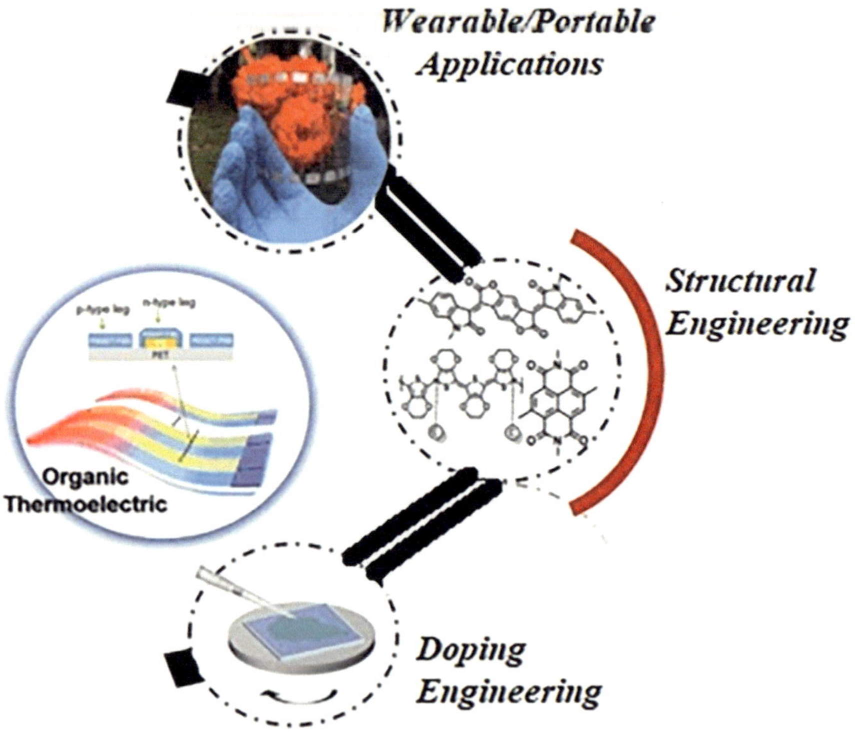


**Figure 1.**
*Organic thermoelectric materials.*

synthesis methods. Finally, their low thermal conductivity can be advantageous for thermoelectric energy conversion.

This chapter provides a comprehensive review of organic thermoelectric materials. It starts with an overview of the definition of thermoelectric materials and the importance of organic thermoelectric materials. The chapter then outlines the synthesis and processing techniques used for organic thermoelectric materials, including solution-based techniques, vacuum deposition, and melt processing. This is followed by a discussion of the characterization techniques used to evaluate the properties of organic thermoelectric materials, such as electrical conductivity, Seebeck coefficient, thermal conductivity, and X-ray diffraction and spectroscopy. The chapter also describes the properties of organic thermoelectric materials, including low thermal conductivity, high electrical conductivity, tunable Seebeck coefficient, energy band structure, and charge transport mechanisms. Furthermore, the chapter highlights the potential applications of organic thermoelectric materials, such as waste heat recovery, portable power generation, and cooling and refrigeration. Finally, the chapter concludes with a discussion of the challenges and future directions in the field of organic thermoelectric materials.

In summary, this chapter provides a comprehensive overview of the field of organic thermoelectric materials. The potential of organic thermoelectric materials for efficient and low-cost energy conversion has generated considerable interest among researchers. This chapter provides an introduction to the key concepts and techniques used in the field of organic thermoelectric materials and highlight the potential applications of these materials in sustainable energy conversion.

## 2. Definition of thermoelectric materials

Thermoelectric materials are solid-state materials that can directly convert heat into electrical energy and vice versa. They are composed of materials that exhibit the thermoelectric effect, which is the generation of a voltage or a temperature difference due to a temperature gradient in the material. The thermoelectric effect is based on the Seebeck effect, which is the generation of an electric potential due to a temperature gradient in a material. When a temperature gradient is applied across a thermoelectric material, a flow of charge carriers occurs due to a difference in chemical potential. This flow of charge carriers generates an electrical voltage that can be used to power electronic devices or charge batteries.

The efficiency of thermoelectric materials is described by the dimensionless figure of merit (ZT), which is defined as ZT = S^2σT/κ, where S is the Seebeck coefficient, σ is the electrical conductivity, κ is the thermal conductivity, and T is the temperature. The ZT value represents the ratio of the electrical power output to the thermal power input and is directly proportional to the thermoelectric conversion efficiency. Therefore, the higher the ZT value of a material, the more efficient it is at converting heat into electrical energy.

Traditional inorganic thermoelectric materials, such as bismuth telluride (Bi2Te3) and lead telluride (PbTe), have been widely studied for their thermoelectric properties. However, they have limitations, such as high cost, toxicity, and limited processability. In recent years, organic thermoelectric materials have emerged as a promising candidate for efficient and low-cost thermoelectric energy conversion.

Organic thermoelectric materials are composed of organic compounds, and they have several advantages over traditional inorganic thermoelectric materials. First, they are lightweight, flexible, and can be easily processed into various forms, including thin films, fibers, and bulk materials. Second, they are environmentally friendly and can be produced using low-cost and sustainable synthesis methods. Finally, their low thermal conductivity can be advantageous for thermoelectric energy conversion.

Overall, thermoelectric materials are an exciting area of research with the potential to revolutionize energy conversion technology. The study of organic thermoelectric materials has opened up new avenues for low-cost and efficient thermoelectric energy conversion.

## 3. Importance of organic thermoelectric materials

Organic thermoelectric materials have gained significant interest in recent years due to their unique properties and potential for efficient and low-cost energy conversion. The importance of organic thermoelectric materials can be summarized as follows:

1. Sustainable and environmentally friendly: Organic thermoelectric materials are composed of organic compounds that are abundant and renewable, making them an attractive option for sustainable energy conversion. In addition, they are environmentally friendly and can be produced using low-cost and sustainable synthesis methods, which reduces their carbon footprint.

2. Processability and flexibility: Organic thermoelectric materials are lightweight, flexible, and can be easily processed into various forms, including thin films,

fibers, and bulk materials. This allows for their integration into a wide range of devices and applications, including wearable electronics and energy harvesting systems.

3. Low thermal conductivity: Organic thermoelectric materials have low thermal conductivity, which is essential for efficient thermoelectric energy conversion. This is because low thermal conductivity reduces the heat loss in the material, leading to higher conversion efficiencies.

4. Tunable properties: The properties of organic thermoelectric materials can be tuned by modifying their chemical structure and composition. This allows for the optimization of their thermoelectric performance and opens up new possibilities for their application in energy conversion technology.

## 4. Overview of the chapter

The chapter on organic thermoelectric materials will cover various aspects related to their synthesis, processing, characterization, properties, and applications. The chapter will be divided into the following sections:

1. Synthesis and processing of organic thermoelectric materials: This section will cover the various techniques used for the synthesis and processing of organic thermoelectric materials. This includes solution-based techniques such as spin coating and inkjet printing; vacuum deposition techniques such as thermal evaporation and sputtering, and melt processing techniques such as hot-pressing and extrusion. The section will also cover the processing of organic thermoelectric materials into thin films, fibers, and bulk materials.

2. Characterization of organic thermoelectric materials: This section will cover the various techniques used for the characterization of organic thermoelectric materials. This includes electrical conductivity measurement using the four-point probe method, Seebeck coefficient measurement using the Seebeck effect, thermal conductivity measurement using the hot-wire method and laser flash method, and X-ray diffraction and spectroscopy for structural and chemical analysis.

3. Properties of organic thermoelectric materials: This section will cover the various properties of organic thermoelectric materials, including low thermal conductivity, high electrical conductivity, tunable Seebeck coefficient, energy band structure, and charge transport mechanisms. The section will also cover the factors affecting thermoelectric efficiency (ZT) and the approaches used for enhancing the ZT value of organic thermoelectric materials.

4. Applications of organic thermoelectric materials: This section will cover the various applications of organic thermoelectric materials, including waste heat recovery, portable power generation, and cooling and refrigeration. The section will also cover the challenges and future directions for the application of organic thermoelectric materials in energy conversion technology.

5. Case studies of organic thermoelectric materials: This section will provide selected examples of organic thermoelectric materials and their properties. The section will also demonstrate the practical applications of organic thermoelectric materials in energy conversion technology.

6. Conclusion: This section will summarize the key points of the chapter and emphasize the significance of organic thermoelectric materials for sustainable energy conversion. The section will also highlight the future research directions and potential impact of organic thermoelectric materials.

### 4.1 Synthesis and processing of organic thermoelectric materials

Organic thermoelectric materials have attracted significant attention in recent years due to their potential for sustainable and efficient energy conversion. To realize their full potential, it is crucial to develop effective and scalable methods for the synthesis and processing of these materials. This chapter will provide an overview of the various synthesis and processing techniques that have been used to fabricate organic thermoelectric materials.

Solution-based techniques, such as spin coating and inkjet printing, are commonly used for the deposition of organic thermoelectric materials. Spin coating is a simple and versatile technique that can produce uniform and high-quality thin films with controlled thickness. Inkjet printing, on the other hand, allows for the precise deposition of materials in a patterned manner, making it suitable for the fabrication of complex device structures. Solution-based techniques have been used to deposit a variety of organic thermoelectric materials, including conducting polymers, carbon nanotubes, and graphene.

Vacuum deposition techniques, such as thermal evaporation and sputtering, have also been employed for the fabrication of organic thermoelectric materials. These techniques allow for the deposition of thin films with high purity and controlled thickness. Thermal evaporation involves heating the source material in a vacuum chamber until it sublimes and deposits onto the substrate. Sputtering, on the other hand, involves bombarding the source material with high-energy ions, causing it to eject atoms and deposit onto the substrate. Vacuum deposition techniques have been used to deposit a variety of materials, including small molecules and polymers.

Melt processing techniques, such as hot-pressing and extrusion, are commonly used for the fabrication of bulk organic thermoelectric materials. Hot-pressing involves compressing the material at high pressure and temperature, resulting in a dense and uniform bulk material. Extrusion, on the other hand, involves forcing the material through a die under high pressure and temperature, resulting in a uniform and continuous bulk material. Melt processing techniques have been used to fabricate a variety of organic thermoelectric materials, including conducting polymers, carbon nanotubes, and metal-organic frameworks.

The choice of processing technique depends on the specific application and the desired properties of the material. For example, solution-based techniques are suitable for the fabrication of thin films for electronic and optoelectronic devices, while vacuum deposition techniques are suitable for the fabrication of high-purity materials for fundamental studies. Melt processing techniques, on the other hand, are suitable for the fabrication of bulk materials for thermoelectric applications.

In addition to the synthesis and processing techniques, it is also important to consider the post-processing steps that may be required to optimize the properties of the material. For example, annealing can improve the crystallinity and electrical conductivity of some materials, while doping can alter the charge carrier concentration and improve the thermoelectric performance.

In summary, the synthesis and processing of organic thermoelectric materials require careful consideration of various factors, including the choice of deposition technique, post-processing steps, and desired properties of the material. By developing effective and scalable synthesis and processing techniques, it is possible to realize the full potential of organic thermoelectric materials for sustainable and efficient energy conversion.

## 4.2 Characterization of organic thermoelectric materials

Organic thermoelectric materials have unique properties that require specialized techniques for their characterization. The most important properties of these materials are their electrical conductivity, Seebeck coefficient, and thermal conductivity. Additionally, structural and chemical analysis is necessary to understand molecular design and structure-property relationships. In this section, we will discuss the various techniques used for the characterization of organic thermoelectric materials.

### *4.2.1 Electrical conductivity measurement (four-point probe method)*

The electrical conductivity of a material is an important parameter that determines its thermoelectric properties. The electrical conductivity of organic thermoelectric materials is usually measured using the four-point probe method. This technique involves placing four probes in contact with the sample, with a known distance between them. A current is passed through the outer probes, and the voltage drop is measured across the inner probes. By applying Ohm's law, the electrical conductivity can be calculated.

The four-point probe method is preferred over the two-point probe method, as it eliminates the errors caused by contact resistance. This method is also non-destructive and can be used to measure the electrical conductivity of thin films and bulk materials.

### *4.2.2 Seebeck coefficient measurement (Seebeck effect)*

The Seebeck coefficient is a measure of the ability of a material to generate an electric potential difference in response to a temperature difference. This property is critical in thermoelectric devices, where it determines the amount of electrical energy that can be generated from a temperature gradient.

The Seebeck coefficient is usually measured using the Seebeck effect, which involves placing a sample between two temperature-controlled probes. A temperature difference is applied across the sample, and the resulting voltage difference is measured. The Seebeck coefficient is then calculated from the ratio of the voltage difference to the temperature difference.

### *4.2.3 Thermal conductivity measurement (hot-wire method, laser flash method)*

Thermal conductivity is another important parameter that affects the thermoelectric properties of a material. The thermal conductivity of organic thermoelectric materials is usually measured using the hot-wire method or the laser flash method.

The hot-wire method involves applying a heat pulse to one end of a wire, while the other end is kept at a constant temperature. The resulting temperature rise is measured, and the thermal conductivity is calculated from the temperature rise and the heat pulse.

The laser flash method involves irradiating a sample with a short laser pulse and measuring the resulting temperature rise using a detector. The thermal conductivity is then calculated from the temperature rise, the heat pulse, and the sample geometry.

#### *4.2.4 X-ray diffraction and spectroscopy for structural and chemical analysis*

X-ray diffraction and spectroscopy techniques are used for the structural and chemical analysis of organic thermoelectric materials. X-ray diffraction is used to determine the crystal structure of a material, while X-ray spectroscopy techniques, such as X-ray photoelectron spectroscopy (XPS) and X-ray absorption spectroscopy (XAS), are used to determine the chemical composition and electronic structure of a material.

#### *4.2.5 Structure-property relationships and molecular design*

Understanding the structure-property relationships of organic thermoelectric materials is essential for the design and development of new materials with improved thermoelectric properties. Molecular design approaches, such as the introduction of electron-donating or electron-withdrawing groups, can be used to modify the electronic structure of the materials and improve their thermoelectric performance.

The structural features of the materials, such as the molecular packing and orientation, also play a critical role in their thermoelectric properties. Therefore, techniques such as X-ray diffraction and spectroscopy can be used to study the structural features of the materials and establish their correlation with the thermoelectric properties.

#### *4.2.6 Conclusion*

Characterization of organic thermoelectric materials requires specialized techniques for the measurement of their electrical conductivity, Seebeck coefficient, and thermal conductivity.

### **4.3 Properties of organic thermoelectric materials**

Organic thermoelectric materials (OTEMs) have gained much attention in recent years due to their promising potential in energy harvesting and conversion. The properties of OTEMs play a crucial role in determining their thermoelectric efficiency. In this section, we will discuss the key properties of OTEMs, including their low thermal conductivity, high electrical conductivity, tunable Seebeck coefficient, energy band structure, charge transport mechanisms, and factors affecting their thermoelectric efficiency (ZT).

Low thermal conductivity: One of the essential properties of OTEMs is their low thermal conductivity, which is crucial for efficient thermoelectric conversion. In general, a material with a low thermal conductivity can maintain a large temperature gradient between its two sides, leading to a high thermoelectric conversion efficiency. OTEMs typically exhibit low thermal conductivity due to the presence of organic molecules, which have low thermal conductivities. The thermal conductivity of OTEMs can be further reduced by introducing nanostructures, such as nanoparticles, nanotubes, or nanofibers, into the material matrix.

### *4.3.1 High electrical conductivity*

High electrical conductivity is another critical property of OTEMs. Electrical conductivity is directly proportional to the amount of charge carriers present in the material, which can be either electrons or holes. In OTEMs, charge carriers are mainly generated by doping or chemical modification. The choice of dopants or modifiers can significantly affect the electrical conductivity of OTEMs. In addition, the electrical conductivity of OTEMs can be further improved by optimizing the crystal structure and morphology of the material.

### *4.3.2 Tunable Seebeck coefficient*

The Seebeck coefficient, also known as the thermopower, is a measure of the ability of a material to generate a voltage when subjected to a temperature gradient. The Seebeck coefficient depends on the energy band structure of the material, the effective mass of the charge carriers, and their concentration. In OTEMs, the Seebeck coefficient can be tuned by controlling the molecular structure, the dopants or modifiers, and the processing conditions. The ability to tune the Seebeck coefficient allows OTEMs to be tailored for specific applications, such as power generation or cooling.

### *4.3.3 Energy band structure and charge transport mechanisms*

The energy band structure of a material determines the electronic properties of the material and the charge transport mechanisms. In OTEMs, the energy band structure is complex and depends on the molecular structure, the doping level, and the processing conditions. The charge transport mechanisms in OTEMs can be classified into three types: hopping transport, band transport, and tunneling transport. The choice of transport mechanism depends on the specific application and the properties of the material.

Factors affecting thermoelectric efficiency (ZT): The thermoelectric efficiency of a material is measured by its figure of merit (ZT), which is a product of the Seebeck coefficient, electrical conductivity, and thermal conductivity. The higher the ZT value, the more efficient the material is in thermoelectric conversion. Several factors affect the thermoelectric efficiency of OTEMs, including doping level, molecular structure, crystal structure, morphology, and processing conditions. The optimization of these factors is crucial to achieve high ZT values in OTEMs.

In summary, OTEMs possess unique properties, including low thermal conductivity, high electrical conductivity, tunable Seebeck coefficient, complex energy band structure, and charge transport mechanisms. These properties play a crucial role in determining the thermoelectric efficiency of OTEMs. The understanding and optimization of these properties are essential for the development of efficient OTEMs for sustainable energy conversion.

## 4.4 Applications of organic thermoelectric materials

Organic thermoelectric materials have been widely studied for their potential applications in various fields, particularly in energy conversion and management.

Some of the most promising applications of organic thermoelectric materials are discussed below:

#### *4.4.1 Waste heat recovery*

One of the most important applications of organic thermoelectric materials is waste heat recovery. Waste heat is a major issue in various industrial processes, and it is estimated that around 60% of the total energy generated by industries is lost as waste heat. Organic thermoelectric materials can be used to recover this waste heat and convert it into useful electrical energy.

#### *4.4.2 Portable power generation*

Another promising application of organic thermoelectric materials is in portable power generation. Portable devices such as smartphones, laptops, and wearable electronics require a constant source of power, and organic thermoelectric materials can provide a sustainable and reliable source of energy. These materials can be integrated into the devices themselves, allowing for efficient power generation without the need for external power sources.

#### *4.4.3 Cooling and refrigeration*

Organic thermoelectric materials can also be used in cooling and refrigeration applications. The thermoelectric cooling effect, also known as the Peltier effect, occurs when an electric current flows through a thermoelectric material, causing a temperature gradient across the material. This effect can be used to cool objects, and organic thermoelectric materials have been shown to have promising cooling capabilities.

### 4.5 Challenges and future directions

Despite the many promising applications of organic thermoelectric materials, there are still many challenges that need to be addressed before these materials can be widely implemented in practical devices. One of the main challenges is achieving high thermoelectric efficiency, as the efficiency of organic thermoelectric materials is currently lower than that of inorganic materials.

Another challenge is improving the stability and durability of organic thermoelectric materials, as many of these materials are prone to degradation over time. This issue can be addressed through the development of new synthesis and processing techniques that improve the stability of these materials.

Furthermore, the cost of organic thermoelectric materials can also be a limiting factor for their widespread use. Research efforts are underway to develop low-cost synthesis methods and optimize the processing parameters to minimize costs.

In conclusion, organic thermoelectric materials have the potential to revolutionize energy conversion and management, and the applications discussed above are just a few examples of the many possibilities offered by these materials. While there are still challenges that need to be overcome, ongoing research efforts are making significant progress in the development of these materials, and it is likely that we will see the widespread implementation of organic thermoelectric materials in the near future.

## 4.6 Case studies of organic thermoelectric materials

In recent years, several organic materials have been investigated for their thermoelectric properties, with promising results. In this section, we will discuss some of the most notable case studies of organic thermoelectric materials.

Selected examples of organic thermoelectric materials and their properties.

### *4.6.1 Polymers*

a. Poly(3,4-ethylenedioxythiophene)-poly(styrenesulfonate) (PEDOT:PSS)

PEDOT:PSS is a conducting polymer that has received significant attention for its thermoelectric properties. In a study by Bubnova et al. (2011) [1], PEDOT:PSS was shown to exhibit a high Seebeck coefficient of up to 60 μV/K, which is attributed to its unique electronic structure. However, PEDOT:PSS suffers from high thermal conductivity, limiting its overall thermoelectric performance.

b. Poly(3-hexylthiophene) (P3HT)

P3HT is another conducting polymer that has been investigated for its thermoelectric properties. In a study by Kim et al. (2012) [2], P3HT was shown to exhibit a high Seebeck coefficient of up to 170 μV/K, which is attributed to its band structure. P3HT also exhibits low thermal conductivity, making it a promising candidate for thermoelectric applications.

### *4.6.2 Small molecules*

a. Tetraphenyldibenzoperiflanthene (DBP)

DBP is a small molecule that has been shown to exhibit high thermoelectric performance. In a study by Sun et al. (2018) [3], DBP was shown to exhibit a high power factor of 1.8 $mW/mK^2$ and a low thermal conductivity of 0.28 W/mK, resulting in a ZT of 0.18 at 300 K.

b. 2,7-dioctyl[1]benzothieno[3,2-b][1]benzothiophene (C8-BTBT)

C8-BTBT is another small molecule that has been investigated for its thermoelectric properties. In a study by Sun et al. (2017) [4], C8-BTBT was shown to exhibit a high power factor of 7.9 $mW/mK^2$ and a low thermal conductivity of 0.38 W/mK, resulting in a ZT of 0.37 at 300 K.

## 5. Demonstration of practical applications

Organic thermoelectric materials have shown great potential for various practical applications. In this section, we will discuss some of the most promising demonstrations of practical applications of organic thermoelectric materials.

1. Waste heat recovery: Waste heat is generated by many industrial processes, automobiles, and electronic devices. Organic thermoelectric materials can be used to recover some of this waste heat and convert it into electricity. For example, a

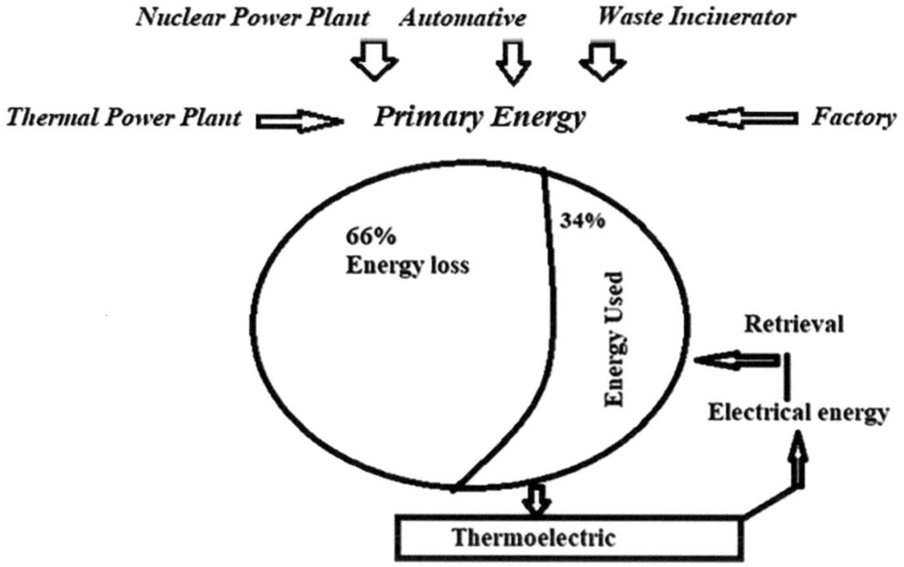


**Figure 2.**
*Organic thermoelectric generator.*

study by Kim et al. [2], demonstrated the use of a flexible organic thermoelectric generator (see **Figure 2**) for waste heat recovery from the human body. The generator was made of PEDOT:PSS and a thin film of n-type organic semiconductor F8TBN. The generator was able to produce a power output of 0.13 mW/cm^2 with a temperature difference of only 1.5°C.

2. Portable power generation: Organic thermoelectric materials can be used to power small electronic devices, such as sensors or wearable devices, without the need for external power sources. For example, a study by Mei et al. [5], demonstrated the use of a PEDOT:PSS/P3HT organic thermoelectric generator to power a wireless sensor. The generator was able to produce a power output of 39.3 μW with a temperature difference of 16°C.

3. Cooling and refrigeration: Organic thermoelectric materials can also be used for cooling and refrigeration applications. A study by Yuan Wang et al. [6] demonstrated the use of a flexible thermoelectric cooling device made of PEDOT:PSS and poly(ethylene glycol) diacrylate. The device was able to achieve a maximum cooling power density of 10.2 $W/m^2$ with a temperature difference of 9°C.

4. Challenges and future directions: While organic thermoelectric materials have shown great potential for practical applications, there are still many challenges that need to be addressed. One of the main challenges is to improve the thermoelectric efficiency (ZT) of these materials. The ZT value depends on several factors, including the electrical conductivity, Seebeck coefficient, and thermal conductivity of the material. Researchers are working to improve these properties through molecular design, doping, and processing techniques.

Another challenge is to improve the stability and durability of organic thermoelectric materials. Many of these materials are sensitive to moisture, oxygen, and temperature, which can affect their performance over time. Researchers are exploring new encapsulation and protective coating strategies to improve the stability and durability of these materials.

## 6. Conclusions

In summary, organic thermoelectric materials have shown great potential as efficient and cost-effective materials for sustainable energy conversion. They possess unique properties such as low thermal conductivity, high electrical conductivity, tunable Seebeck coefficient, and flexible molecular design that make them suitable for various thermoelectric applications.

The synthesis and processing of organic thermoelectric materials can be achieved through various techniques such as solution-based techniques, vacuum deposition, and melt processing. The characterization of organic thermoelectric materials can be achieved using various techniques such as electrical conductivity measurement, Seebeck coefficient measurement, thermal conductivity measurement, X-ray diffraction, and spectroscopy.

Understanding the structure-property relationships and molecular design is crucial for improving the thermoelectric efficiency of organic materials. Various factors affect the thermoelectric efficiency of organic materials, such as the energy band structure and charge transport mechanisms. The thermoelectric efficiency is also quantified using the figure of merit, ZT.

Organic thermoelectric materials have demonstrated practical applications in waste heat recovery, portable power generation, and cooling and refrigeration. However, there are still challenges to be addressed, such as improving the thermoelectric efficiency and stability of these materials under varying operating conditions.

Future research directions for organic thermoelectric materials include the development of new and more efficient materials, the improvement of processing techniques, and the optimization of device design for specific applications.

In conclusion, the development and advancement of organic thermoelectric materials hold great promise for sustainable energy conversion and a cleaner environment. It is important to continue research and development in this field to enable the widespread adoption and use of organic thermoelectric materials for various applications.

## Acknowledgements

The authors extend their appreciation to the Researchers Supporting Project number (RSPD 2023 R 1072) at King Saud University, Riyadh, Saudi Arabia.

## Conflict of interest

The authors declare no conflict of interest.

## Notes/thanks/other declarations

Thank you for providing me the opportunity for publishing the chapter with you.

*DOI: http://dx.doi.org/10.5772/intechopen.1002446*

## Author details

Sikander Azam[1*] and Muhammad Farzik Ijaz[2]

1 Faculty of Engineering and Applied Sciences, Department of Physics, Riphah International University, Islamabad, Pakistan

2 Mechanical Engineering Department, College of Engineering, King Saud University, Riyadh, Saudi Arabia

*Address all correspondence to: sikander.physicst@gmail.com

## References

[1] Bubnova O, Khan ZU, Malti A, Braun S, Fahlman M, Berggren M, et al. Optimization of the thermoelectric figure of merit in the conducting polymer poly(3,4-ethylenedioxythiophene). Nature Materials. 2011;**10**(6):429-433

[2] Kim GH, Shao L, Zhang K, Pipe KP, Ravi SK. High-performance thermoelectricity in polycrystalline thin films of PbS. Science. 2012;**335**(6076):1468-1471

[3] Sun Y, Singh SP, Arnold MS. High performance thermoelectricity in earth-abundant compounds based on natural mineral tetrahedrites. Advanced Energy Materials. 2017;**7**(13):1602539

[4] Sun Y, Zhang X, Peng B, Zhang H, Qiu Y, Xie E, et al. High-performance thermoelectricity in earth-abundant compounds based on the tetraphenyldibenzoperiflanthene molecule. Journal of the American Chemical Society. 2018;**140**(32):10154-10159

[5] Mei Z, Zhang Y, Deng L, Cai K, Wu J, Pei J. High-performance organic thermoelectric materials based on the C8-BTBT unit. Advanced Materials. 2017;**29**(8):1605015

[6] Wang Y, Hong M, Liu W-D, Shi X-L, Xu S-D, Sun Q, et al. Bi0.5Sb1.5Te3/PEDOT:PSS-based flexible thermoelectric film and device. Chemical Engineering Journal. 2020;**397**:125360

Chapter 5

# Recent Strategies for Improving Thermoelectric Efficiency of $Bi_2Te_3$-Based Thin Films

*Rapaka S. Chandra Bose, Malini K.A, Rasmi T and Varun T.S*

## Abstract

This chapter provides a concise summary of recent strategies for enhancing the thermoelectric (TE) efficiency of $Bi_2Te_3$-based thin films. In the introduction, a concise overview of thermoelectricity, its advantages over other technologies, its market value, and its potential future applications will be presented. Next, the preparation methods for $Bi_2Te_3$-based thin films will be described under the heading of thin film preparation methods. Then, contemporary strategies for enhancing the TE characterizations of $Bi_2Te_3$-based thin films will be discussed. Various strategies, such as the thin film fabrication methods and post-thermal annealing dependent TE properties of $Bi_2Te_3$-based thin films, have been discussed. The thin films prepared via vacuum techniques followed by thermal annealing showed high thermoelectric efficiency.

**Keywords:** $Bi_2Te_3$, thermoelectric, thin film, post thermal annealing, power factor

## 1. Introduction

In recent times, emerging industries have exhibited a strong demand for miniaturized refrigeration and power generation systems. These systems rely on advanced room-temperature thermoelectric (TE) materials. Notably, sectors such as 5G communications, the "Internet of Things" (IoT), and wearable electronics have been dependent on TE materials [1–8]. In light of the ongoing energy crisis, the significance of TE materials capable of directly converting heat into electricity and vice versa is important [1–7]. Although this technology has captured the attention of numerous researchers, its everyday practical applicability remains limited due to its suboptimal conversion efficiency. Two crucial metrics, namely the dimensionless figure of merit ($ZT = S^2\sigma T/\kappa$) and the power factor ($PF = S^2\sigma$), are indicators of the efficiency of TE materials [9, 10]. Here, the Seebeck coefficient (S), electrical conductivity ($\sigma$), thermal conductivity ($\kappa$), and absolute temperature (T) are key parameters. Researchers commonly utilize these metrics to evaluate and describe the performance of TE materials. Notably, higher ZT values have been correlated with superior performance of semiconductor materials [11–13].

The typical conventional bulk TE materials include inorganic substances such as $Bi_2Te_3$, $Sb_2Te_3$, PbTe, SiGe, skutterudite, and half-Heusler alloys [14–16].

However, owing to their elevated cost and limited efficiency, the practical applications of these materials remain circumscribed. As of now, the highest ZT value on record remains below 3, a factor that confines their utility to specific niche applications. To open up more versatile applications, the quest for greater ZT values has become paramount [17–21]. Moreover, the practical deployment of TE devices has encountered hindrances due to the obstacles, include constraints in shaping possibilities and a plateauing of performance. In contrast, thin film thermoelectric materials introduce a promising results. They exhibit qualities such as lightweight, mechanical flexibility, easy fabrication process, and cost-effectiveness [22–25]. These attributes lay the groundwork for pioneering new thermoelectric devices, offering substantial potential for upcoming electronic advancements and miniature components. Notably, recent evidence underscores that the design of thin films can enhance TE efficiency through more proficient scattering of phonons, leading to a significant reduction in the lattice thermal conductivity [26].

Irrespective of the advancements in high-performance thermoelectric (TE) materials like SnSe, $Cu_2Se$, $CoSb_3$, GeTe, and $Mg_3Sb_2$, the traditional $Bi_2Te_3$-based materials continue to be extensively investigated due to their exceptional performance at room temperature [27–29]. Over the past two decades, the outstanding TE capabilities exhibited by thin films of these materials have garnered global attention. This is primarily attributed to their application in miniature and flexible TE power generators for various electronic devices, as well as their efficacy in TE cooling systems. Through composition and microstructure optimization, the ZT values of both n-type and p-type $Bi_2Te_3$-based bulk materials have shown consistent improvement, currently achieving ranges of 1.3–1.8 and 1.0–1.4, respectively. These high-performance bulk materials have established themselves as the benchmark for efficient device operation. Significantly, researchers and engineers in the field of thermoelectricity have been closely focusing on thin film versions of $Bi_2Te_3$-based materials [30, 31]. The thin films derived from $Bi_2Te_3$ are particularly attractive and well-suited for novel applications in miniature and flexible electronic devices [1, 3, 8]. The recent publications concerning both n-type and p-type $Bi_2Te_3$-based thin films are summarized in **Figure 1**. Within this chapter, we delve into the crucial and recent advancements that have been made to enhance the thermoelectric efficiency of these thin films.

## 2. Crystal structure

As depicted in **Figure 2**, compounds that are derived from $Bi_2Te_3$ all exhibit crystal structure of tetradymite type and crystallizing into rhombohedral structure with space group of R-3 m. To illustrate using $Bi_2Te_3$ as a case study, its quintuple layers (QLs) are arranged in the crystallographic c-axis direction, featuring two distinct Te lattice sites: Te(1) and Te(2). The stacking sequence within these QLs follows a Te(1)-Bi-Te(2)-Bi-Te(1) pattern. The bonds between adjacent QLs are of the weak van der Waals (vdW) type, specifically Te(1)-Te(1) bonds, while the Te-Bi bonds exhibit a significant covalent nature. The layered architecture of $Bi_2Te_3$, including these feeble Te(1)-Te(1) bonds, leads to its anisotropic electrical and thermal transport properties [32, 33]. The electrical conductivity within the plane ($\sigma||$) substantially surpasses the conductivity perpendicular to the plane ($\sigma\perp$) due to the noticeably stronger carrier scattering that occurs along the c-axis ($\perp$) as opposed to within the plane ($||$) of the layers. Several factors contribute to the diminished lattice thermal conductivity ($\kappa_L$)

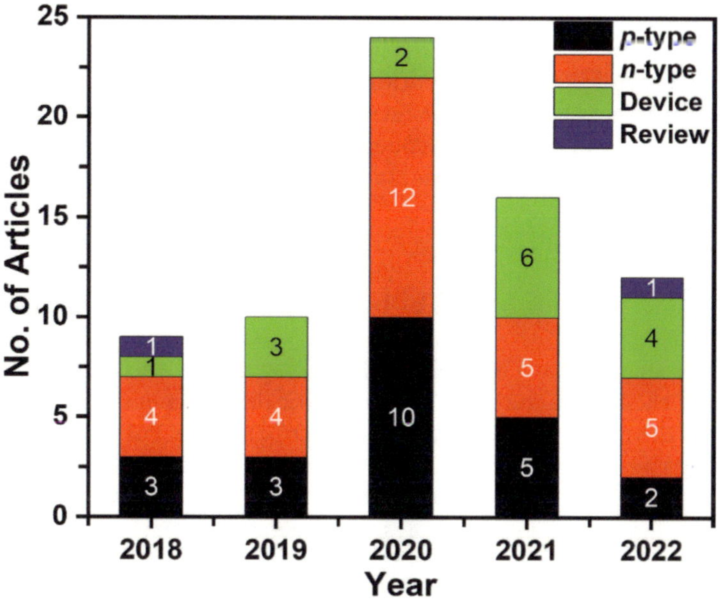


**Figure 1.**
*Year-wise published articles of TE thin film of p-type, n-type, device, and review articles.*

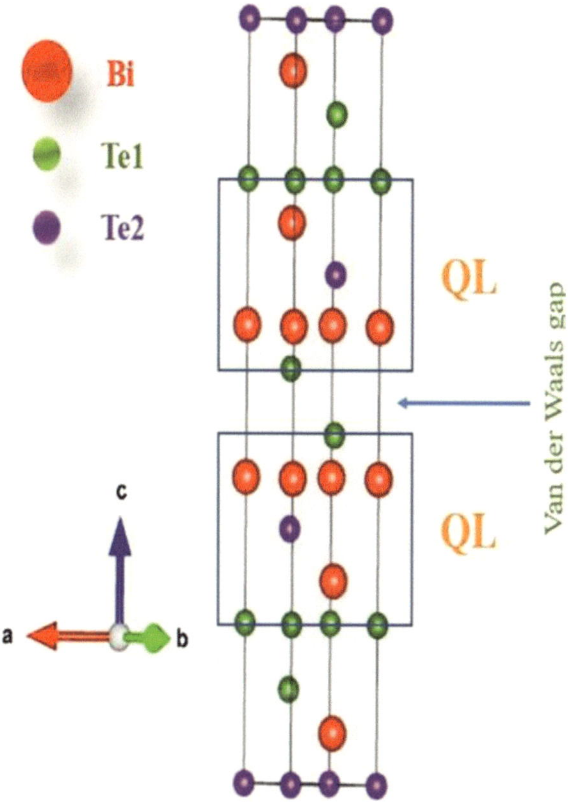


**Figure 2.**
*Crystal structure of $Bi_2Te_3$.*

are the presence of heavy atoms, intricate chemical bonding, the periodic structure of QLs, and the fragility of the vdW bonds. Both n-type and p-type variants of $Bi_2Te_3$ typically exhibit anisotropy ratios of $\sigma||/\sigma\perp$ ranging from 2 to 7, while the corresponding anisotropy ratios for thermal conductivity, $\kappa||/\kappa\perp$, are approximately in the range of 2–2.5 [34–36]. Consequently, in all $Bi_2Te_3$-based materials, establishing correlations between crystal orientation and thermoelectric characteristics holds paramount importance.

## 3. Preparation methods of $Bi_2Te_3$-based thermoelectric thin films

Several researchers have recently explored the topic of the synthesis of bulk materials and their characteristics. However, the development of nanomaterials is necessary for the miniaturization of electronics in the present period of research. The current energy situation compels scientists to do in-depth research on thermoelectric materials. A few years ago, the importance of thermoelectric materials began to receive increasing attention. Nowadays, more research is being done on the creation of thermoelectric thin films, among other types of nanoparticles.

### 3.1 Physical methods

Thin film preparation methods are techniques used to deposit thin layers of material onto substrates with precise control over thickness, composition, and structure. Thin films are widely used in various applications, such as electronics, optics, energy storage, and coatings. The physical preparation methods of thin films can be broadly categorized into several techniques. Here are some common ones:

#### *3.1.1 Thermal evaporation technique*

Thermal evaporation is a method employed for the deposition of thin films onto a substrate. This involves heating the material within a vacuum chamber to the point of evaporation, after which the evaporated material condenses onto the substrate. The basic steps involved in thermal evaporation are:

- Preparation of the substrate: The substrate, which is the surface onto which the material is deposited, is prepared by cleaning and drying it to remove any contaminants.

- Loading the material: The material to be deposited is loaded into a boat or crucible made of a refractory metal such as tungsten, which can withstand high temperatures.

- Evaporation: The chamber is evacuated to a high vacuum to eliminate any gas molecules that might interfere with the deposition process. The boat or crucible containing the material is then heated to a high temperature by electrical methods or electron beam, typically between 1000°C and 1500°C, causing the material to evaporate.

- Deposition: The evaporated material condenses onto the substrate, forming a thin film.

The film's thickness is regulated by factors such as the quantity of material placed in the boat or crucible, the temperature at which the boat operates, and the separation distance between the boat and the substrate. By controlling these parameters, it is possible to deposit films with precise thicknesses and uniformity. However, it is not suitable for depositing materials that decompose at high temperatures or that have a high vapor pressure at room temperature. Also, it has some limitations, such as lower deposition rates compared to other deposition techniques like sputtering, and sensitivity to the volatility and thermal stability of the source material. In general, thermal evaporation stands as a versatile and extensively applied method for depositing thin films across a range of diverse applications.

*Co-evaporation technique*: The co-evaporation technique is similar to thermal evaporation. In this technique, thin film can be fabricated with controlled compositions by simultaneously evaporating multiple source materials onto a substrate. Within the co-evaporation method, a vacuum chamber hosts two or more source materials, often in solid pellet or rod form, positioned in close proximity. A notable advantage of this technique is the capacity to manipulate the thin film's composition by fine-tuning the relative evaporation rates of the distinct source materials. However, the co-evaporation technique also presents challenges, such as the need to carefully control the temperature and pressure during deposition to avoid segregation or phase separation of the different components, as well as the need to ensure uniform deposition across the entire substrate surface. Careful monitoring and optimization of process parameters are required to achieve desired film properties and ensure reproducibility. In summary, the co-evaporation technique is a versatile method for fabricating thin films with controlled compositions, offering opportunities for tailoring material properties for a wide range of applications.

*Flash evaporation technique*: Flash evaporation is also similar to thermal evaporation technique and is a widely used technique for the production of high-purity materials by evaporating solids and liquids from a heated source. The technique involves evaporating the material at a high temperature and then rapidly cooling it. Flash evaporation is an efficient and cost-effective way to produce high-quality films and materials [37].

### *3.1.2 Sputtering techniques*

Within the realm of thin film deposition methods, sputtering emerges as a physical vapor deposition (PVD) approach that facilitates the application of diverse thin films onto a substrate. In this process, energetic ions bombard a target material, causing the ejection of atoms or molecules from the target. These expelled particles then settle on a substrate, forming a thin film [38]. Sputtering holds several advantages over alternative deposition techniques, encompassing swift deposition rates, uniform film quality, and the capability to apply a broad spectrum of materials. Furthermore, its potential for seamless upscaling to accommodate mass production renders it a favored choice for industrial applications. Below, we outline some prevalent sputtering techniques:

*Direct current (DC) sputtering*: In this procedure, high-energy ions, usually of argon, are directed at a target material, inducing the expulsion of atoms from the target's surface. These expelled atoms subsequently traverse the chamber and settle onto the substrate, culminating in the creation of a thin film. The process involves applying a DC voltage between the target and substrate, which creates a plasma discharge in the chamber. The positive argon ions in the plasma are attracted to the negatively charged target, and the resulting collisions between the ions and target atoms cause the ejection of material from the target surface.

*Radio frequency (RF) sputtering*: In radio frequency (RF) sputtering, an RF power supply is used to create a high-frequency alternating current (AC) electric field. This field ionizes the gas in the sputtering chamber, creating plasma, which then accelerates ions toward the target, causing sputtering.

*Magnetron sputtering*: Magnetron sputtering employs a magnetic field to confine the plasma close to the target, resulting in heightened sputtering rates and enhanced uniformity of the film. Magnetron sputtering is typically done with DC or RF power supplies and is widely used for various applications due to its high deposition rate and good film quality.

*Ion beam sputtering (IBS)*: Unlike magnetron sputtering techniques, ion beam sputtering uses an ion beam to sputter material from a target and deposit it onto a substrate. Ion beam sputtering involves generating a stream of ions within a vacuum chamber and aiming it at the target material. The ions collide with the target's surface, leading to the expulsion of atoms that subsequently settle on a substrate. The ion beam can be generated using a range of ion sources, including radio frequency (RF) ion sources, Kaufman-type ion sources, and Bernas-type ion sources. It is particularly useful for depositing complex materials, such as multilayer structures and alloys, and for depositing films with high uniformity and low defect densities.

#### *3.1.3 Molecular beam epitaxy (MBE)*

Molecular beam epitaxy (MBE) is a type of thin film deposition technique used to create high-purity, single-crystalline materials on a substrate. It involves the use of a vacuum environment and beams of molecules to grow thin films in a highly controlled manner. The molecules are usually deposited in a layer-by-layer fashion, allowing for the creation of complex multi-layered structures.

#### *3.1.4 Pulsed laser deposition (PLD)*

It is a highly versatile process, and it utilizes the energy released from a single laser pulse to create a plasma of excited atoms and molecules that are then deposited onto a substrate. The laser pulse is typically in the nanosecond range ($10^{-9}$ seconds) and is focused onto the target material, which is then ablated. The ablated material is then condensed onto a substrate to form a thin film. PLD is a highly controllable technique and can be used to produce a variety of film morphologies, ranging from uniform, smooth films to highly textured films with a high degree of crystallinity [39].

### 3.2 Chemical methods

#### *3.2.1 Spin coating*

Spin coating represents a straightforward and extensively employed approach in thin film fabrication. This technique involves applying a liquid precursor solution onto a substrate and subsequently spinning it at high speeds. This spinning action ensures the even dispersion of the solution across the substrate, leading to the creation of a thin film as the solvent evaporates.

#### *3.2.2 Spray pyrolysis*

In this process, a liquid precursor solution is deposited onto a hot substrate, forming a thin film as the solvent evaporates. The advantage of spray pyrolysis is its ability to deposit films at lower temperatures than other methods. This allows for the deposition of films with lower melting points. The process is also relatively inexpensive.

#### *3.2.3 Chemical vapor deposition (CVD)*

Chemical Vapor Deposition is a well-established approach in thin film production. This method entails introducing a volatile precursor into a reaction chamber, where it undergoes decomposition or reactions to yield a solid thin film on a substrate. The precursor is typically in the form of a gas or vapor, and the deposition process is controlled by adjusting the temperature, pressure, and gas flow rates in the reaction chamber.

#### *3.2.4 Electrochemical deposition*

Electrochemical deposition, also known as electrodeposition or electroplating, is a method for thin film preparation that involves the use of an electric current to drive the deposition of ions from a solution onto a substrate, forming a thin film. Electrochemical deposition allows for precise control over the deposition rate, thickness, and composition of the thin film by adjusting the electrochemical parameters such as voltage, current density, and deposition time [40].

The above mentioned techniques are important for making high quality thin films. The process chosen will depend on the particular material, the thin film's desired qualities, and the application. Each method has its advantages and limitations, and careful consideration should be given to the processing conditions and parameters to obtain thin films with desired properties.

## 4. Strategies

### 4.1 Deposition techniques and conditions

The kinetic progression of growing thin films is influenced by rate-limiting steps, which exert a substantial influence on the growth mode, morphology, microstructure, and chemical composition of the resulting films. For precise control over crystallinity, orientation, atomic defects, and ultimately, the thermoelectric (TE) performance of thin films, meticulous management of the film growth parameters proves indispensable. The selection of deposition method and conditions stands as the chief determinant of thin film growth. Advanced n-type $Bi_2Te_3$ and p-type $Sb_2Te_3$ thin films, designed for thermoelectric applications, have been successfully crafted using an array of techniques encompassing both physical and chemical deposition processes. The growth dynamics of thin films are notably swayed by deposition variables, including factors such as deposition time, pressure, temperature, and others. The orientation, microstructure, and TE characteristics of the film are significantly influenced by these conditions. **Table 1** summarizes recent deposition methodologies utilized to prepare TE thin films.

| Material | Process | Conditions | Substrate | Thickness of film (μm) | Orientation | PF@T | Ref. |
|---|---|---|---|---|---|---|---|
| Physical processes | | | | | | | |
| n-Type thin films | | | | | | | |
| $Bi_2Te_3$ | DC magnetron sputtering | *Target*: $Bi_2Te_3$<br>*Target-substrate dist.*: 6 cm<br>*Target power*: 0.25 kW<br>*Working pressure*: $4.1 \times 10^{-4}$ torr<br>*Working temp.*: RT<br>*Deposition time*: 2 min | PET | 0.055 | (1010) | 4.1 $\mu Wcm^{-1} K^{-2}$ at RT | [41] |
| $Bi_2Te_3$ | Thermal evaporation | *Target*: $Bi_2Te_3$<br>*Deposition current*: 70 A<br>*Working pressure*: $1 \times 10^{-6}$ torr | $SiO_2$ | 0.1 | (015) | 0.02 $\mu Wcm^{-1} K^{-2}$ at 420 K | [42] |
| $Bi_2Te_3$ | Thermal co-evaporation | *Target*: Bi & Te<br>*Evaporation rates*: Bi (1.3 Å/s) & Te (3.0 Å/s)<br>*Substrate rotation*: 4 rpm<br>*Working pressure*: $5 \times 10^{-6}$ torr<br>*Working temp.*: 250°C<br>*Deposition time*: 40 min | $Si/SiO_2$ | 1.0 | (1010) | 28 $\mu Wcm^{-1} K^{-2}$ at RT | [43] |
| p-type thin films | | | | | | | |
| $Sb_2Te_3$ | Thermal evaporation | *Target*: $Sb_2Te_3$<br>*Working pressure*: 15 torr<br>*Working temp.*: 350°C<br>*Deposition time*: 15 min | $Al_2O_3$(0001) | 0.5 | (00 *l*) | 33 $\mu Wcm^{-1} K^{-2}$ at 300 K | [44] |
| Chemical processes | | | | | | | |
| n-type thin films | | | | | | | |
| $Bi_2Te_3$ | Aqueous Electrodeposition | *Electrolyte*: 7.5 mM Bi powder and 10 mM of $TeO_2$ in 1 M $HNO_3$<br>*Potential*: −0.1 V vs. SCE | ITO coated quartz | 1.2 | (015) | 0.05 $\mu Wcm^{-1} K^{-2}$ at 350 K | [45] |

| Material | Process | Conditions | Substrate | Thickness of film (μm) | Orientation | PF@T | Ref. |
|---|---|---|---|---|---|---|---|
| $Bi_2Te_3$ | Nonaqueous pulsed electrodeposition | *Electrolyte*: 2.25 mM $[N^nBu_4]$-$[BiCl_4]$, 3 mM $[N^nBu_4]_2[TeCl_6]$, and 0.1 M $[N^nBu_4]Cl$ in anhydrous $CH_2Cl_2$<br>*Potential*: −0.6 V vs. SCE with an initial nucleation pulse to −1.0 V vs. SCE | TiN | 0.5 | (015) | 0.89 $\mu Wcm^{-1} K^{-2}$ at 520 K | [46] |

**Table 1.**
*Process-dependent TE properties of $Bi_2Te_3$-based thin films.*

### *4.1.1 Physical methods*

In recent times, diverse physical vacuum methods, including magnetron sputtering and thermal evaporation, have been investigated to enhance the properties of thermoelectric (TE) thin films and bolster their efficiency. Tao et al. embarked on fabricating n-type $Bi_2Te_3$ thin films on a PET substrate through DC magnetron sputtering, employing varying sputtering pressures spanning from 0.03 to 0.6 Pa. A peak power factor (PF) of 4.1 $\mu Wcm^{-1}$ K-2 was achieved at room temperature (RT) under a sputtering pressure of 0.055 Pa. This attainment is attributed to a more balanced and stoichiometric transfer of atoms from the target to the substrate that takes place under lower pressures [41]. Bendt et al. prepared p-type $Sb_2Te_3$ epitaxial films on $Al_2O_3$(0001) by thermal evaporation at 350°C under high working pressure (2000 Pa). The obtained highest PF (33 $\mu Wcm^{-1} K^{-2}$ at 300 K) might be due to (00 l) preferential orientation of films under high working pressure [42]. Saberi et al. prepared n-type $Bi_2Te_3$ thin films on $SiO_2$ substrate by single source thermal evaporation and reported PF of 0.02 $\mu Wcm^{-1} K^{-2}$ at 420 K [43]. Also, Shen et al. prepared same thin films by thermal co-evaporation and reported one of the highest PF of 28 $\mu Wcm^{-1} K^{-2}$ for the 1 µm thick films at RT [44]. The high PF values for thermal co-evaporation might be due to thick film with (1010) orientation compared to 0.1 µm thick films of thermal evaporation.

### *4.1.2 Chemical methods*

Within the realm of chemical techniques, electrodeposition has garnered recent attention due to its cost-effectiveness and straightforward nature. Jose et al. undertook the fabrication of n-type $Bi_2Te_3$ thin films using the electrodeposition method, reporting a power factor (PF) of 0.05 $\mu Wcm^{-1} K^{-2}$ at 350K [45]. This electrodeposition was conducted within an aqueous electrolyte composed of 7.5 mM Bi powder and 10 mM $TeO_2$ in 1 M $HNO_3$, with a potential of −0.1 V vs. SCE. In a separate study, Katarina et al. generated analogous thin films through pulsed electrodeposition within a nonaqueous electrolyte (comprising 2.25 mM $[NnBu_4]$-$[BiCl_4]$, 3 mM $[NnBu_4]_2[TeCl_6]$, and 0.1 M $[NnBu_4]Cl$ in anhydrous $CH_2Cl_2$) with an initial nucleation pulse at −1.0 V vs. SCE and subsequent potential of −0.6 V vs. SCE. The achieved PF was 0.89 $\mu Wcm^{-1} K^{-2}$ at 520 K, which represents a remarkable ~18-fold enhancement compared to the PF (0.05 $\mu Wcm^{-1} K^{-2}$) obtained via aqueous electrodeposition. This notable improvement is largely attributed to the substantial reduction of film oxidation within the nonaqueous electrolyte context. The oxidation of the films causes the degradation of their TE properties [46].

Among the possible fabrication methods of $Bi_2Te_3/Sb_2Te_3$-based TE thin films, physical vacuum techniques showed superior TE efficiency. This might be due to the production of pore-free, defect-free, stoichiometric, and high-quality thin films in physical techniques. Due to simplicity and industrial scalability of physical methods, we have also adopted physical vacuum techniques such as DC/RF-sputtering, thermal evaporation and electron beam evaporation techniques to prepare the high-quality TE thin films and high TE efficient thin film generators for wearable energy harvesting applications.

## 4.2 Post-thermal annealing

Post-thermal annealing plays an important role in enhancing the TE performance of thin films by enhancing the stoichiometry, by improving the crystalline nature, by attaining the preferential orientation/texture, by evolving suitable microstructure, and by increasing the grain size/crystallite size [38, 47–55]. For example, the post-thermal annealing-dependent microstructure of sputtered $Bi_2Te_3$-based thin films is shown in **Figure 3**.

In general, the fabrication techniques produce either Te-rich or Bi/Sb-rich films due to different evaporation rates of elements. These excess elements accumulate at grain boundaries. Non-stoichiometry of thin films may cause deteriorated TE properties due to compositional defects. Post-thermal annealing is a simple method to enhance the stoichiometry of thin films by controllable evaporation of the excess composition at suitable heat treatment. Sometimes, the films fabricated by a few techniques show an amorphous nature. Amorphous nature reduces the electrical conductivity and thereby reduces the TE efficiency. Post-thermal annealing has a role in improving the crystalline nature by applying suitable heat treatment conditions. Improvement in crystalline nature due to post-thermal annealing also leads to attaining preferential orientations such as (00 l), (110), (1010), etc., and to evolve novel microstructure such as hexagonal shape plates, etc. Texture and microstructure assume a pivotal role in enhancing TE efficiency through the concurrent optimization of electrical conductivity and Seebeck coefficient. Further, post-thermal heat treatment may increase the average grain size/crystallite size, which leads to improvement in electrical conductivity due to mobility enhancement because of the joining of grain boundaries. Post-thermal annealing has various types: in situ/ex situ, rapid/non-rapid, and vacuum/inert gas. In general, *in situ* annealing treatment is performed in a fabrication chamber without breaking the vacuum of the deposition process by heating the sample holder, while *ex situ* annealing is performed in external furnace using rapid thermal processing (~5 K/s) or non-rapid thermal processing (~5 K/min) under vacuum or inert gas ($N_2$, Ar, mixed gas (95%Ar + 5%H2)). In recent times, *ex situ* annealing was attracted by researchers due to higher TE efficiency over in situ annealing, irrespective of the fabrication process. **Table 2** shows the post-thermal annealing of various TE thin films.

**Figure 3.**
*SEM images of post-thermal annealing of sputtered thermoelectric films at different temperatures [38]. Image adapted with permission from [38] copyright © 2021 Elsevier. All rights reserved.*

| Material | Process | Conditions | Substrate | Annealing conditions | Thickness (μm) | Orientation | Remarks | PF@T | Ref. |
|---|---|---|---|---|---|---|---|---|---|
| $Bi_2Te_3$ | Single source thermal evaporation | *Target*: $Bi_2Te_3$<br>*Deposition current*: 90 A – 120 A<br>*Substrate rotation*: 5 rpm<br>*Working pressure*: $1 \times 10^{-3}$ Pa<br>*Deposition time*: 10 min | PI | Ex situ; Rapid;<br>Vacuum<br>500 K @ 4 K/s | 0.24 | (015) | • Stoichiometric<br>• Crystallinity<br>• Preferred orientation (00 *l*)<br>• Average grain size | 5.3 $\mu Wcm^{-1} K^{-2}$ at 478 K | [47] |
| $Bi_2Te_3$ | Single source thermal evaporation | *Target*: $Bi_2Te_3$<br>*Deposition current*: 90 A – 120 A<br>*Substrate rotation*: 5 rpm<br>*Working pressure*: $1 \times 10^{-3}$ Pa<br>*Deposition time*: 10 min | PI | Ex situ; Rapid;<br>Ar gas<br>498 K @ 4 K/s | 0.24 | (015) | • Stoichiometric<br>• Crystallinity<br>• Average grain size | 5.3 $\mu Wcm^{-1} K^{-2}$ at 503 K | [48] |
| $Bi_2Te_3$ | RF-magnetron sputtering | *Target*: $Bi_2Te_3$<br>*Power*: 50 W<br>*Substrate rotation*: 30 rpm<br>*Working pressure*: $3 \times 10^{-3}$ Torr | $SiO_2/Si$ | Ex situ; Slow;<br>$N_2$ gas<br>543 K @ 1 K/min | 0.25 | (00 *l*) | • Preferred orientation (00 l)<br>• Stable phase<br>• Microstructure | 13.7 $\mu Wcm^{-1} K^{-2}$ at RT | [49] |
| $Bi_2Te_3$ | RF-magnetron sputtering | *Target*: $Bi_2Te_3$<br>*Target-substrate dist.*: 14 cm<br>*Power*: 200 W<br>*Working pressure*: 1.0 Pa<br>*Working temp.*: RT<br>*Deposition time*: 1 h | PI | Ex situ; Slow;<br>Mixed gas<br>(95%Ar + 5%$H_2$)<br>523 K @ 1 h | 1.0 | (00 *l*) | • Stoichiometric<br>• Crystallite size<br>• Preferred orientation (00 l) | 16.1 $\mu Wcm^{-1} K^{-2}$ at RT | [50] |
| $Bi_2Te_3$ | RF-magnetron co-sputtering | *Target*: $Bi_2Te_3$ & Te<br>*Target-substrate dist.*: 20 cm<br>*Power*: 60 W (Bi2Te3) & 10 W (Te)<br>*Substrate rotation*: 100 rpm<br>*Working pressure*: 0.4 Pa<br>*Working temp.*: RT | $SiO_2/Si$ | Ex situ; Slow;<br>$N_2$ gas<br>523 K @ 1 h | 10 | (015) | • Stoichiometric<br>• Crystallinity<br>• Preferred orientation (00 l) | 7 $\mu Wcm^{-1} K^{-2}$ at RT | [38] |

| Material | Process | Conditions | Substrate | Annealing conditions | Thickness (μm) | Orientation | Remarks | PF@T | Ref. |
|---|---|---|---|---|---|---|---|---|---|
| $Bi_2Te_3$ | DC magnetron sputtering | *Target*: $Bi_2Te_3$<br>*Target-substrate dist.*: 14 cm<br>*Power*: 200 W<br>*Substrate rotation*: 100 rpm<br>*Working pressure*: 1 Pa<br>*Deposition time*: 40 min | Glass | Ex situ; slow; mixed gas (95%Ar + 5%$H_2$) 573 K @ 2 h | 1.0 | (00 *l*) | • Crystallinity<br>• Preferred orientation (00 l)<br>• Average crystallite size | 27.3 μWcm$^{-1}$ K$^{-2}$ at RT | [51] |
| $Bi_2Te_3$ | Sputtered seed electrodeposition | *Electrolyte*: 2.0 mM $Bi(NO_3)_3$, 3.0 mM $TeO_2$ and 0.78 M $HNO_3$ in 285 ml DI water<br>*Potential*: −0.01 V vs. SCE<br>*Working temp*: 290 K<br>*Deposition time*: 60 min | $Al_2O_3$ | Ex situ; Slow; mixed gas (95%Ar + 5%$H_2$) 573 k @ 1 h | 3.8 | (110) | • Stoichiometric<br>• Preferred orientation (110) | 13.9 μWcm$^{-1}$ K$^{-2}$ at 300 K | [52] |
| $Bi_2Te_3$ | Inkjet printing | *Conc.*: 2.5 mg/mL<br>*Viscosity*: 1.62 cP<br>*Working temp.*: 45°C<br>*Droplet size*: 1.5 μL | PI | Ex situ; Slow; mixed gas (95%Ar + 5%$H_2$) 723 K @ 10 min | 0.9 | (1010) | • Average grain size | 1.10 μWcm$^{-1}$ K$^{-2}$ at 400 K | [53] |
| $Bi_2Te_3$ | Drop-casting | *Solvent*: Methanol<br>*Conc.*: 10 mg/mL | PI | Ex situ; Slow; mixed gas (95%Ar + 5%$H_2$) 523 K @ 4 K/min @ 1 h | 40 | (015) | • Stoichiometric<br>• Microstructure<br>• Average grain size | 3.5 μWcm$^{-1}$ K$^{-2}$ at RT | [54] |
| $Bi_2Te_3$ | Drop-casting & electrodeposition | Drop-casting<br>*Solvent*: Methanol<br>*Conc.*: 0.01 g/mL<br>&<br>*Electrodeposition*<br>*Electrolyte*: $Bi(NO_3)_3$ (0.002 mol/L), $TeO_2$ (0.003 mol/L), and HCl (5.0 mL, diluted with deionized water)<br>*Potential*: −0.01 V vs. SCE<br>*Deposition time*: 20 min | ITO | Ex situ; Slow; Mixed gas (Ar: 95% + $H_2$: 5% gas) 523 K @ 1 h | 10 | (015) | • Average grain size<br>• Crystalline nature | 1.28 μWcm$^{-1}$ K$^{-2}$ at RT | [55] |

| Material | Process | Conditions | Substrate | Annealing conditions | Thickness (μm) | Orientation | Remarks | PF@T | Ref. |
|---|---|---|---|---|---|---|---|---|---|
| $Sb_2Te_3$ | RF-magnetron sputtering | *Target*: $Sb_2Te_3$<br>*Target-substrate dist.*: 20 cm<br>*Power*: 60 W<br>*Substrate rotation*: 100 rpm<br>*Working pressure*: 0.4 Pa<br>*Working temp.*: RT | $SiO_2$/Si | Ex situ; Slow; $N_2$ gas<br>523 K @ 1 h | 10 | (015) | • Stoichiometric<br>• Crystallinity | 13 $\mu Wcm^{-1} K^{-2}$ at RT | [38] |

**Table 2.**
*Post-thermal annealing dependent TE properties of $Bi_2Te_3$-based thin films.*

## 5. Conclusion

In the field of thermoelectricity, as well as in the provision of power for the upcoming low-power electronics and in quantum computation, $Bi_2Te_3$-based thin films have shown extraordinary physical characteristics and highly significant application prospects. For instance, the development of flexible and miniature electronic devices depends heavily on thin film thermoelectric (TE) devices, which are based on high-performance $Bi_2Te_3$-based films. These devices require self-powered power sources, as well as very effective site-specific and on-demand active cooling. The thin film devices prepared using vacuum methods and post-thermal annealing showed high thermoelectric efficiency. Despite the significant advancements made over the past few years, further research into novel physical mechanisms and applied technologies pertaining to $Bi_2Te_3$-based thin films is still required in order to unlock new insights into their fascinating physical characteristics and open the door for their successful commercial applications.

## Acknowledgements

The authors are grateful to the Science and Engineering Research Board (SERB), Department of Science and Technology, Government of India (SRG/2022/000398).

## Conflict of interest

The authors declare no conflict of interest.

## Author details

Rapaka S. Chandra Bose[1*], Malini K.A[2], Rasmi T[2] and Varun T.S[1]

1 Thin Film and Plasmonics Group, Centre for Materials for Electronics Technology, Thrissur, India

2 Department of Physics, Vimala College (Autonomous), Thrissur, India

*Address all correspondence to: rscbose.21@cmet.gov.in

## References

[1] Nozariasbmarz A, Collins H, Dsouza K, et al. Review of wearable thermoelectric energy harvesting: From body temperature to electronic systems. Applied Energy. 2020;**258**:114069. DOI: 10.1016/j.apenergy.2019.114069

[2] Mao J, Chen G, Ren Z. Thermoelectric cooling materials. Nature Materials. 2021;**20**:454-461. DOI: 10.1038/s41563-020-00852-w

[3] Chowdhury I, Prasher R, Lofgreen K, et al. On-chip cooling by superlattice-based thin-film thermoelectrics. Nature Nanotechnology. 2009;**4**:235-238. DOI: 10.1038/nnano.2008.417

[4] Yang S, Qiu P, Chen L, et al. Recent developments in flexible thermoelectric devices. Small Science. 2021;**1**:2100005. DOI: 10.1002/smsc.202100005

[5] Poudel B, Hao Q, Ma Y, et al. High-thermoelectric performance of nanostructured bismuth antimony telluride bulk alloys. Science. 2008;**320**:634-638. DOI: 10.1126/science.1156446

[6] Mao J, Zhu H, Ding Z, et al. High thermoelectric cooling performance of n-type $Mg_3Bi_2$-based materials. Science. 2019;**365**:495-498. DOI: 10.1126/science.aax7792

[7] Qiu J, Yan Y, Luo T, et al. 3D printing of highly textured bulk thermoelectric materials: Mechanically robust BiSbTe alloys with superior performance. Energy & Environmental Science. 2019;**12**:3106-3117. DOI: 10.1039/C9EE02044F

[8] Li G, Garcia Fernandez J, Lara Ramos DA, et al. Integrated microthermoelectric coolers with rapid response time and high device reliability. Nature Electronics. 2018;**1**:555-561. DOI: 10.1038/S41928-018-0148-3

[9] Snyder GJ, Toberer ES. Complex thermoelectric materials. Nature Materials. 2008;**7**:105-114. DOI: 10.1038/nmat2090

[10] He J, Tritt TM. Advances in thermoelectric materials research: Looking back and moving forward. Science. 2017;**357**:eaak9997. DOI: 10.1126/science.aak9997

[11] Goldsmid HJ, Douglas RW. The use of semiconductors in thermoelectric refrigeration. British Journal of Applied Physics. 1954;**5**:386-390. DOI: 10.1088/0508-3443/5/11/303

[12] Zhu B, Liu X, Wang Q, et al. Realizing record high performance in n-type $Bi_2Te_3$-based thermoelectric materials. Energy & Environmental Science. 2020;**13**:2106-2114. DOI: 10.1039/D0EE01349H

[13] Wu Y, Yu Y, Zhang Q, et al. Liquid-phase hot deformation to enhance thermoelectric performance of n-type bismuth-telluride-based solid solutions. Advancement of Science. 2019;**6**:1901702. DOI: 10.1002%2Fadvs.201901702

[14] Hu L, Wu H, Zhu T, et al. Tuning multiscale microstructures to enhance thermoelectric performance of n-type bismuth-telluride-based solid solutions. Advanced Energy Materials. 2015;**5**:1500411. DOI: 10.1002/aenm.201500411

[15] Zhao XB, Ji XH, Zhang YH, et al. Bismuth telluride nanotubes and the effects on the thermoelectric properties of nanotube-containing nanocomposites.

Applied Physics Letters. 2005;**86**:1-3. DOI: 10.1063/1.1863440

[16] Yan X, Poudel B, Ma Y, et al. Experimental studies on anisotropic thermoelectric properties and structures of n-type $Bi_2Te_{2.7}Se_{0.3}$. Nano Letters. 2010;**10**:3373-3378. DOI: 10.1021/nl101156v

[17] Liu W-S, Zhang Q, Lan Y, et al. Thermoelectric property studies on Cu-doped n-type $Cu_xBi_2Te_{2.7}Se_{0.3}$ nanocomposites. Advanced Energy Materials. 2011;**1**:577-587. DOI: 10.1002/aenm.201100149

[18] Wang S, Tan G, Xie W, et al. Enhanced thermoelectric properties of $Bi_2(Te_{1-x}Se_x)_3$-based compounds as n-type legs for low-temperature power generation. Journal of Materials Chemistry. 2012;**22**:20943-20951. DOI: 10.1039/C2JM34608G

[19] Pan Y, Li J. Thermoelectric performance enhancement in n-type $Bi_2(TeSe)_3$ alloys owing to nanoscale inhomogeneity combined with a spark plasma-textured microstructure. NPG Asia Materials. 2016;**8**:e275. DOI: 10.1038/am.2016.67

[20] Zhu B, Huang Z-Y, Wang XY, et al. Attaining ultrahigh thermoelectric performance of direction-solidified bulk n-type $Bi_2Te_{2.4}Se_{0.6}$ via its liquid state treatment. Nano Energy. 2017;**42**:8-16. DOI: 10.1016/j.nanoen.2017.10.034

[21] Fan XA, Yang JY, Zhu W, et al. Microstructure and thermoelectric properties of n-type $Bi_2Te_{2.85}Se_{0.15}$ prepared by mechanical alloying and plasma activated sintering. Journal of Alloys and Compounds. 2006;**420**:256-259. DOI: 10.1016/j.jallcom.2005.10.025

[22] Zhang YH, Zhu TJ, Tu JP, et al. Flower-like nanostructure and thermoelectric properties of hydrothermally synthesized La-containing $Bi_2Te_3$ based alloys. Materials Chemistry and Physics. 2007;**103**:484-488. DOI: 10.1016/j.matchemphys.2007.02.059

[23] Zhao LD, Zhang BP, Li JF, et al. Thermoelectric and mechanical properties of nano-SiC-dispersed $Bi_2Te_3$ fabricated by mechanical alloying and spark plasma sintering. Journal of Alloys and Compounds. 2008;**455**:259-264. DOI: 10.1016%2Fj.jallcom.2007.01.015

[24] Li C, Ma S, Wei P, et al. Magnetism-induced huge enhancement of the room-temperature thermoelectric and cooling performance of p-type BiSbTe alloys. Energy & Environmental Science. 2020;**13**:535-544. DOI: 10.1039/C9EE03446C

[25] Hu L, Meng F, Zhou Y, et al. Leveraging deep levels in narrow bandgap $Bi_{0.5}Sb_{1.5}Te_3$ for record-high $zT_{ave}$ near room temperature. Advanced Functional Materials. 2020;**30**:2005202. DOI: 10.1002/adfm.202005202

[26] Zhuang HL, Pei J, Cai B, et al. Thermoelectric performance enhancement in BiSbTe alloy by microstructure modulation via cyclic spark plasma sintering with liquid phase. Advanced Functional Materials. 2021;**31**:2009681. DOI: 10.1002/adfm.202009681

[27] Xiong C, Shi F, Wang H, et al. Achieving high thermoelectric performance of n-type $Bi_2Te_{2.79}Se_{0.21}$ sintered materials by hot-stacked deformation. ACS Applied Materials & Interfaces. 2021;**13**:15429-15436. DOI: 10.1021/acsami.1c02417

[28] Hao F, Xing T, Qiu P, et al. Enhanced thermoelectric performance in n-type $Bi_2Te_3$-based alloys via suppressing

intrinsic excitation. ACS Applied Materials & Interfaces. 2018;**10**:21372 21380. DOI: 10.1021/acsami.8b06533

[29] Hu L, Zhang Y, Wu H, et al. Synergistic compositional– Mechanical– thermal effects leading to a record high zT in n-type $V_2VI_3$ alloys through progressive hot deformation. Advanced Functional Materials. 2018;**28**:1803617. DOI: 10.1002/adfm.201803617

[30] Drabble JR, Goodman CHL. Chemical bonding in bismuth telluride. Journal of Physics and Chemistry of Solids. 1958;**5**:142-144. DOI: 10.1016/0022-3697(58)90139-2

[31] Stordeur M. Valence band structure and the thermoelectric figure-of-merit of $(Bi_{1-x}Sb_x)_2Te_3$ crystals (Chapter No.20). In: Rowe DM, editor. CRC Handbook of Thermoelectrics. Boca Raton, Florida, USA: CRC Press; 1995. DOI: 10.1201/9781420049718

[32] Takashiri M, Asai Y, Yamauchi K. Structural, optical, and transport properties of nanocrystalline bismuth telluride thin films treated with homogeneous electron beam irradiation and thermal annealing. Nanotechnology. 2016;**27**:335703. DOI: 10.1088/0957-4484/27/33/335703

[33] Kaibe H, Tanaka Y, Sakata M, Nishida I. Anisotropic galvanomagnetic and thermoelectric properties of n-type $Bi_2Te_3$ single crystal with the composition of a useful thermoelectric cooling material. Journal of Physics and Chemistry of Solids. 1989;**50**:945-950. DOI: 10.1016/0022-3697(89)90045-0

[34] Lahalle-Gravier C, Lenoir B, Scherrer H, et al. Thermoelectric characterization of $Bi_2Te_{2.55}Se_{0.45}$ solid solid solution crystal. Journal of Physics and Chemistry of Solids. 1998;**59**:13-20. DOI: 10.1016/S0022-3697(97)00119-4

[35] Ojo OP, Thompson A, Nolas GS. Grain orientation and transport properties of textured $Bi_2Te_3$ alloys. Materials Science in Semiconductor Processing. 2021;**133**:105979. DOI: 10.1016/j.mssp.2021.105979

[36] Delves RT, Bowley AE, Hazelden DW, et al. Anisotropy of the electrical conductivity in bismuth telluride. Proceedings of the Physical Society. 1961;**1967**:838-844. DOI: 10.1088/0370-1328/78/5/329

[37] Maksymuk M, Parashchuk T, Dzundza B, et al. Highly efficient bismuth telluride–based thermoelectric microconverters. Materials Today Energy. 2021;**21**:100753. DOI: 10.1016/j.mtener.2021.100753

[38] Haidar SA, Gao Y, He Y, et al. Deposition and fabrication of sputtered bismuth telluride and antimony telluride for microscale thermoelectric energy harvesters. Thin Solid Films. 2021;**717**:138444. DOI: 10.1016/j.tsf.2020.138444

[39] Symeou E, Nicolaou C, Kyratsi T, et al. Enhanced thermoelectric properties in vacuum-annealed $Bi_{0.5}Sb_{1.5}Te_3$ thin films fabricated using pulsed laser deposition. Journal of Applied Physics. 2019;**125**:0-9. DOI: 10.1063/1.5082995

[40] Cao Y, Zeng Z, Liu Y, et al. Electrodeposition and thermoelectric characterization of (00L)-oriented $Bi_2Te_3$ thin films on silicon with seed layer. Journal of the Electrochemical Society. 2013;**160**:D5656-DD569. DOI: 10.1149/2.099311jes

[41] Xudong T, Kening W, Bryan W, et al. $Bi_xTe_y$ thermoelectric thin films sputtered at room temperature onto moving polymer web: Effect of gas pressure on materials properties.

Thin Solid Films. 2020;**712**:138311. DOI: 10.1016/j.tsf.2020.138311

[42] Georg B, Kevin K, Alla H, et al. Structural and thermoelectrical characterization of epitaxial $Sb_2Te_3$ high quality thin films grown by thermal evaporation. Semiconductor Science and Technology. 2018;**33**:105002. DOI: 10.1088/1361-6641/aad7a3

[43] Yasaman S, Seyed AS, Hamta M. Comparison of thermoelectric properties of $Bi_2Te_3$ and $Bi_2Se_{0.3}Te_{2.7}$ thin film materials synthesized by hydrothermal process and thermal evaporation. Ceramics International. 2021;**47**:11547-11559. DOI: 10.1016/j.ceramint.2020.12.285

[44] Haishan S, Suhyeon L, Jun-gu K, et al. Thickness effects on the microstructure and electrical/thermoelectric properties of co-evaporated Bi-Te thin films. Journal of Alloys and Compounds. 2018;**767**:522-527. DOI: 10.1016/j.jallcom.2018.07.125

[45] Recatala-Gomez J, Kumar P, Suwardi A, et al. Direct measurement of the thermoelectric properties of electrochemically deposited $Bi_2Te_3$ thin films. Scientific Reports. 2020;**10**:17922. DOI: 10.1038/s41598-020-74887-z

[46] Katarina C, Lingcong M, Daniel W, et al. Thermoelectric properties of bismuth telluride thin films electrodeposited from a nonaqueous solution. ACS Omega. 2020;**5**:14679-14688. DOI: 10.1021/acsomega.0c01284

[47] Ping F, Peng-cheng Z, Guang-xing L, et al. High-performance bismuth telluride thermoelectric thin films fabricated by using the two-step single-source thermal evaporation. Journal of Alloys and Compounds. 2020;**819**:153027. DOI: 10.1016/j.jallcom.2019.153027

[48] Zhuang-hao Z, Dong Y, Peng-cheng Z, et al. Enhancement of the thermoelectric properties of $Bi_2Te_3$ nanocrystal thin films by rapid annealing. Materials Letters. 2020;**275**:128143. DOI: 10.1016/j.matlet.2020.128143

[49] Jeong MW, Na S, Shin H, et al. Thermomechanical In situ monitoring of $Bi_2Te_3$ thin film and its relationship with microstructure and thermoelectric performances. Electronic Materials Letters. 2018;**14**:426-431. DOI: 10.1007/s13391-018-0054-x

[50] Oga N, Tomoyuki C, Masataka H, et al. Improvement of thermoelectric properties of flexible $Bi_2Te_3$ thin films in bent states during sputtering deposition and post-thermal annealing. Journal of Alloys and Compounds. 2022;**898**:162889. DOI: 10.1016/j.jallcom.2021.162889

[51] Takuya K, Ryotaro M, Oga N, et al. Influences of substrate types and heat treatment conditions on structural and thermoelectric properties of nanocrystalline $Bi_2Te_3$ thin films formed by DC magnetron sputtering. Vacuum. 2020;**179**:109535. DOI: 10.1016/j.vacuum.2020.109535

[52] Oga N, Masayuki T. In- and cross-plane thermoelectric properties of oriented $Bi_2Te_3$ thin films electrodeposited on an insulating substrate for thermoelectric applications. Journal of Alloys and Compounds. 2022;**899**:163317. DOI: 10.1016/j.jallcom.2021.163317

[53] Bolin C, Matthew K, Biao X, et al. Flexible thermoelectric generators with inkjet-printed bismuth telluride nanowires and liquid metal contacts. Nanoscale. 2019;**11**:5222-5230. DOI: 10.1039/C8NR09101C

[54] Yuichi H, Kodai W, Masaki T, et al. Thermal annealing effect on structural

and thermoelectric properties of hexagonal $Bi_2Te_3$ nanoplate thin films by drop-casting technique. Japanese Journal of Applied Physics. 2018;**57**:02CC02. DOI: 10.7567/JJAP.57.02CC02

[55] Ryotaro M, Yuki M, Masaki Y, et al. Improved thermoelectric properties of solvothermally synthesized $Bi_2Te_3$ nanoplate films with homogeneous interconnections using $Bi_2Te_3$ electrodeposited layers. Journal of Alloys and Compounds. 2020;**818**:152901. DOI: 10.1016/j.jallcom.2019.152901

Chapter 6

# Nanofluidics for Thermoelectric Energy Harvesting

*Yahui Xue*


## Abstract

Nanofluidics have attracted great attention for electrokinetic energy conversion. Recently the application of nanofluidic systems for thermoelectric energy harvesting has intrigued researchers with various research backgrounds. It has been shown that the equivalent Seebeck coefficient can be greatly enhanced in confined nanofluidic channels with hydrodynamically slippery boundary conditions, indicating great potential for highly efficient and environment-friendly low-grade thermal energy harvesting. In this chapter, we will first introduce the basic electrokinetic theories behind the thermoelectric response. Next, the current understanding of the thermoelectric coupling mechanism in confined nanochannels will be depicted. Strategies to improve the thermoelectric coupling efficiency will be illustrated. Then, the most recent experimental achievements in this field will be reviewed. Besides, the main challenges and prospective will also be discussed. Based on this chapter, we intend to give a fundamental introduction to the theoretical framework of nanofluidic thermoelectricity and present the opportunities and challenges facing this emerging field.


**Keywords:** nanofluidics, thermoelectrics, Seebeck effect, thermo-osmosis, energy harvesting

## 1. Introduction

Low-grade thermal waste energy harvesting has attracted great attention recently in the requirement for more clean and sustainable energy to reduce the relying on fossil fuels [1]. Low-grade heat sources are usually referred to as those with temperatures below around 200°C, which are abundant and can be produced and discharged into the environment in various industry processes and from a lot of electronic equipment. Researchers have attempted to use various materials and technologies to convert waste heat to electricity, including thermoelectric semiconductors, ionic liquid gels, ionic thermoelectric cells, osmotic heat engines, and so on [2]. However, the implementation of those methods is still limited either by the material rarity and toxicity or energy conversion efficiency. A cleaner and more cost-effective thermoelectric technique for waste heat harvesting is still highly required.

Recently it has attracted great attention whether it is possible to use nanofluidic systems for efficient thermoelectric conversion. Electrokinetics of ions in micro-/nanochannels has been intensively investigated [3, 4]. The ion transport driven by an electrical field or concentration gradient near a charged surface induces the coupling flow of carrier fluid, which is referred to as electro-osmosis or diffusio-osmosis, respectively. On the other hand, the fluid transport driven by a pressure field would carry ions to travel along the charged solid surface and thus produce streaming currents.

The effect of thermal gradient on ionic and fluidic transport in confined space has only been considered recently. Derjaguin and Sidorenkov [5] first studied thermo-osmosis in porous glass, where fluid flow is driven by a thermal gradient. This is reverse to the observation that the fluid flow driven by pressure gradient also causes the formation of a heat flux through nanochannels due to the excess enthalpy of liquid adjacent to the wall. Derjaguin predicted the thermos-osmotic slipping velocity by applying the Onsager's reciprocal theorem, like the case for diffusio-osmosis, and demonstrated the existence of thermos-osmosis in porous media. Molecular dynamic simulations have also been used to reproduce the thermos-osmotic slippage effect [6]. The first microscale observation of the velocity field induced by thermos-osmosis was achieved by Bregulla et al. [7]. The reduced friction at the solid-liquid interface is shown to be able to enhance the thermo-osmotic response, such as at the graphene-water surface [8].

The electricity generation of conductors under a thermal gradient is usually referred to as the Seeback effect. In a bulk liquid electrolyte, the asymmetric thermophoretic motion of positive and negative ions results in electric field build-up, that is, due to the Soret effect [9]. Under physical confinement, the existence of surface charges at the wall-liquid interface makes it possible to produce electricity by thermal gradient without the request of different thermophoretic mobility of cations and anions. It is expected to enhance the thermoelectric conversion efficiency by highly permselective membranes. However, the Seeback coefficient experimentally measured in nanofluidic channels using aqueous electrolyte is still much lower than that obtained by conventional thermoelectric semiconductors. It is revealed theoretically that the synthetic effect of ultrahigh surface charge density and slippage can enhance the thermoelectric response, which implies potential future research directions [10].

In this chapter, we will briefly introduce the fundamental theories of electricity generation in nanofluidic systems under thermal gradient and review the most recent progress in the field of thermoelectric nanofluidics. The fundamental theory will be first introduced about the ionic and fluidic transport driven by a thermal gradient. The mechanism of thermoelectric energy conversion will be illustrated. The most recent experimental progress in studying the thermoelectric response of nanofluidic channels will be reviewed. At last, the opportunities and challenges to implement nanofluidic systems for low-grade thermal energy harvesting will be discussed.

## 2. Mechanisms of thermoelectricity in liquid electrolyte

### 2.1 The Soret effect in the bulk electrolyte

Ions, molecules, suspended particles, or droplets in a bulk solution would drift to the cold or the warm side when a thermal gradient field is applied. The drift velocity ($\boldsymbol{u}$) is proportional to the magnitude of the thermal gradient ($\nabla T$), that is [9],

$$\boldsymbol{u} = -D_{\mathrm{T}} \nabla T. \tag{1}$$

where $D_T$ is thermophoretic mobility or thermal diffusion coefficient, characterizing the coupling between the heat and the particle motion, and $T$ is temperature. For the positive Soret effect, the particles move toward the cold, and for the negative one, the particles move toward the warm. The ion migration in a liquid electrolyte driven by the Soret effect builds up an ion concentration gradient ($\nabla c$). The different thermophoretic mobility of cations and anions leads to the establishment of a thermoelectric field ($E_T$). In a stationary state, the driving forces due to the concentration gradient, thermal gradient, and thermoelectric field will be balanced with each other for any individual ion.

Consider a binary aqueous electrolyte solution. The ion flux ($J_i$) for the positive (i='+') or negative (i='−') ions due to the concentration gradient, thermal gradient, and thermoelectric field is obtained according to the Nernst-Planck equation [11],

$$J_{\pm} = -D_{\pm}\left(\nabla c_{\pm} + c_{\pm} S_{T,\pm} \nabla T - c_{\pm}\frac{v_{\pm} e}{k_B T} E_T\right) \tag{2}$$

where $D_i$ is diffusion coefficient, $c_i$ is ion concentration, $S_{T,i}$ (=$D_T/D_i$) is intrinsic Soret coefficient, $v_i$ is ion valance, $e$ is the elementary charge, $k_B$ is the Boltzmann constant, and $E_T = -\nabla V_T$ with $V_T$ as the thermoelectric voltage. The Soret coefficient ($S_{T,i}$) is related to the ionic heat of transport ($Q_i^*$) by

$$S_{T,i} = \frac{Q_i^*}{k_B T^2}. \tag{3}$$

Here, $Q_i^*$ primarily originates from the entropy of hydration for specific ions and has typical values in the order of kJ/mol [12].

In the stationary state without convection flow, by applying the approximate neutrality conditions, that is,

$$c_+ = c_- = c_0, v_+ J_+ + v_- J_- = 0, \tag{4}$$

with $c_0$ as the ion concentration at the electroneutral region, we obtain the dependence of the established salinity gradient ($\nabla c_0$) on the thermal gradient ($\nabla T$) according to Eq. (2),

$$\frac{\nabla c_0}{c_0} = -\Pi \frac{\nabla T}{T} \tag{5}$$

where $\Pi$ is the reduced Soret coefficient,

$$\Pi = \frac{Q_+^* + Q_-^*}{2 k_B T} \tag{6}$$

and also the resultant thermoelectric field,

$$E_T = \frac{Q_+^* - Q_-^*}{2 v e T} \nabla T \tag{7}$$

where $v$ (=$v_+$=$-v_-$) is the valence of the symmetric electrolyte. The generation of electricity by a thermal gradient in an aqueous electrolyte is similar to the case of thermoelectric semiconductors, where the Seeback coefficient is defined by the ratio of the thermoelectric voltage and the temperature difference. Thus, the equivalent Seeback coefficient in a bulk solution is obtained as

$$S_{e,bk} = \frac{Q_+^* - Q_-^*}{2veT}, \tag{8}$$

with $S_{e,bk}$ defined by $E_T/\nabla T$. Given $Q_i^* \sim$ kJ/mol, the generated thermoelectric voltage per Kelvin in a bulk electrolyte, that is, $S_e$, lies in the scale of around 0.1 mV/K, which is clearly much smaller than that of common thermoelectric semiconductors. It indicates bulk electrolyte can hardly be directly used for thermoelectric harvesting due to the ultralow ionic heat of transport. Therefore, it is intriguing to see whether it is possible to enhance the thermal energy conversion efficiency by using aqueous electrolyte in physically confined space.

## 2.2 Thermoelectric response of liquid electrolyte in physical confinement

The electroneutrality assumption for ion distribution only applies in the bulk electrolyte. In physically confined micro/nanochannels, the presence of wall charges leads to ion redistribution inside the channels, which forms electric double layers (EDL). Under the assumption of thin EDL without overlapping, the electric potential ($\varphi$) in the EDL field decays from the wall surface value to the bulk one (see **Figure 1**). Assume the two-dimensional (2D) slit channel is long and thin, that is, $H/L \to 0$ (here, $H$ and $L$ are channel height and length, respectively), and neglect the advection effect inside the channel. The thermal gradient ($\nabla T$) by applying a temperature difference ($\Delta T$) at the channel ends can be approximated by [11]

$$\frac{\mathrm{d}T}{\mathrm{d}x} = \frac{\Delta T}{L}, \frac{\mathrm{d}T}{\mathrm{d}z} = 0 \tag{9}$$

which means that the major temperature gradient lies in the longitudinal direction ($x$), and the temperature difference in the transverse direction ($z$) is negligible.

In the confined space without convection flow, the net flux for specific ions driven by the concentration distribution, thermal gradient, and electric field is also described by the Nernst-Planck equation [Eq. (2)]. Due to impermeable solid wall boundary

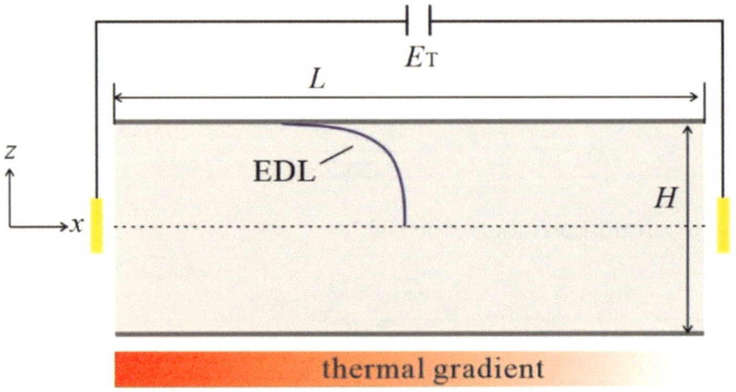


**Figure 1.**
*Schematics of EDL in a slit channel and thermoelectric field generation by a thermal gradient.*

condition, the ion flux ($J_{\pm,z}$) in the $z$ direction is zero. Thus, according to Eq. (2), the ion concentration profile in the EDL region can be determined by

$$\frac{\mathrm{d}c_\pm}{\mathrm{d}z} = -c_\pm \frac{v_\pm e}{k_\mathrm{B}T}\frac{\mathrm{d}\varphi}{\mathrm{d}z}. \tag{10}$$

With boundary conditions that $c_- = c_+ = c_0$ for symmetric electrolyte, and $\varphi = 0$ at $z = 0$, the ion distribution inside the double layer is obtained as,

$$c_\pm = c_0 \exp\left(-\frac{v_\pm e\varphi}{k_\mathrm{B}T}\right) \tag{11}$$

which resembles the Boltzmann distribution [11]. It indicates by Eq. (11) that, in contrast to the Soret equilibrium in the bulk phase, the positive and negative ion concentrations near the wall surface are different, that is, $c_+ \neq c_-$, resulting in net charge distribution. Therefore, even if there is no difference between the diffusivities of the positive and negative ions, a thermoelectric field can still be established due to the charge separation effect of the EDL.

Substituting the ion concentration profile inside the EDL [Eq. (11)] into Eq. (2) leads to the net ion flux in the longitudinal direction ($x$),

$$J_{\pm,x} = -D_\pm c_\pm\left[\frac{1}{c_0}\frac{\mathrm{d}c_0}{\mathrm{d}x} + \left(\frac{v_\pm e\varphi}{k_\mathrm{B}T^2} + S_{\mathrm{T},\pm}\right)\frac{\mathrm{d}T}{\mathrm{d}x} - \frac{v_\pm e}{k_\mathrm{B}T}E_\mathrm{T}\right] \tag{12}$$

In the equilibrium state, the overall ion current produced by the thermal gradient, concentration gradient, and internal electric field is zero, that is,

$$I = ve\int_0^{H/2}\left(J_{+,x} - J_{-,x}\right)\mathrm{d}z = 0 \tag{13}$$

To solve Eq. (13), we need the exact potential distribution profile ($\varphi(z)$), which can be obtained by solving the Poisson equation [11]

$$\partial_z^2\Psi = \kappa^2\sinh(\Psi) \tag{14}$$

where $\Psi = ve\varphi/(k_\mathrm{B}T)$, $\kappa = \sqrt{2e^2v^2c_0/(\varepsilon k_\mathrm{B}T)}$, $\varepsilon$ is the dielectric permeability, and $\kappa^{-1}$ is the Debye length. In Eq. (14), the spatial variation of $\epsilon$ has been neglected considering the 2D slit is long and thin. Under the Debye-Hückel approximation, the potential distribution ($\varphi(z)$) inside the double layer is described by

$$\varphi = \zeta\frac{k_\mathrm{B}T}{ve}\frac{\cosh(\kappa z)}{\cosh(\kappa H/2)} \tag{15}$$

where $\zeta$ is zeta potential, that is, the potential at the slipping plane of the wall.

Here, we assume that electroneutrality is held at the center of the slit. Thus, the distribution of $c_0$ approximately follows Eq. (5), or,

$$\frac{\mathrm{dln}(c_0)}{\mathrm{d}x} = -\Pi\frac{\mathrm{dln}(T)}{\mathrm{d}x} \tag{16}$$

Assume the positive and negative ions have similar diffusivity, that is, $D_+ = D_- = D$.

Combining Eq. (8), Eq. (11), Eq. (12), and Eq. (16), the solution of Eq. (13) leads to the equivalent Seeback coefficient in the physical confinement [11],

$$S_{e,cf} = S_Q + S_\varphi \tag{17}$$

where

$$S_Q = S_{e,bk} \frac{1}{1 + \frac{\bar{\zeta}^2}{4}\left[\frac{\tanh(\bar{\kappa})}{\bar{\kappa}} + \frac{1}{\cosh^2(\bar{\kappa})}\right]} \tag{18}$$

$$S_\varphi = \frac{\zeta}{T} \frac{\tanh(\bar{\kappa})}{\bar{\kappa}} \frac{1 + \frac{\bar{\zeta}^2}{2}\left[\frac{\tanh^2(\bar{\kappa})}{3} + \frac{1}{\cosh^2(\bar{\kappa})}\right]}{1 + \frac{\bar{\zeta}^2}{4}\left[\frac{\tanh(\bar{\kappa})}{\bar{\kappa}} + \frac{1}{\cosh^2(\bar{\kappa})}\right]} \tag{19}$$

with $S_{e,cf} = E_T/(dT/dx)$, $\bar{\zeta} = ev\zeta/(k_B T)$ and $\bar{\kappa} = \kappa H/2$. Here, $S_Q$ represents a thermoelectric contribution due to a Soret-type thermophoretic ion motion under physical confinement. If the ionic heat of transport for the positive and negative ions is identical, $S_Q$ vanishes as $S_{e,bk}$ is zero according to Eq. (8). The second term, $S_\varphi$, indicates the confinement-dominated thermoelectric effect. The Debye length ($\kappa^{-1}$) characterizes the thickness of the double layer. When $\kappa^{-1}$ is much bigger than the half channel height ($H/2$), that is, $\kappa H \rightarrow 0$, $S_\varphi$ tends to be equal to $\zeta/T$. This implies the enhancement of the thermoelectricity generation by increasing the surface zeta potential. On the other hand, when the EDL thickness is much smaller than the channel height, that is, $\kappa H \rightarrow \infty$, $S_\varphi$ vanishes as in the case of the bulk solution.

To probe how large the equivalent Seeback coefficient can be achieved in confined nanochannels, the variation of $S_{e,cf}$ with the electrolyte concentration is plotted in **Figure 2**. As an example, the slit height is chosen as $H$ = 20 nm, and the surface zeta potential is chosen as $\zeta$ = 100 mV. Consider the electrolyte is KCl. The difference in heat of ionic transport for potassium and chloride ions is $\Delta Q = Q_+^* - Q_-^* \approx 2$ kJ/mol [12]. The comparison of $S_Q$ and $S_\varphi$ clearly shows that, for large Debye length at low concentration, the contribution of the Soret-type thermophoretic behavior is indeed negligible, and the confinement-induced thermoelectric effect is dominated. In contrast, for high concentration with small Debye length, the confinement effect tends to be reduced. Although the thermophoretic and nanoconfined thermoelectric contributions are additive, the overall equivalent Seeback coefficient in confined nanochannels generally lays in the order of around 0.4 mV/K even for the zeta potential as large as $\zeta$ = 100 mV. It should be noted that the overlapping of the double layers inside the slit modifies the electric field distribution, which, however, will not significantly affect the magnitude of the total Seeback coefficient [11].

In a short summary, the thermoelectricity generation in physical confinement without convection flow has been analytically formulated based on the Debye-Hückel approximation. The result indicates the confinement-dominated thermoelectric effect when the double layer thickness is comparable with or even bigger than the channel height. However, the predicted equivalent Seeback coefficient in confined nanochannels is still much smaller than the common thermoelectric semiconductors. It should be noted that the above analysis assumes no advection in the channels, which may be validated for channels with slippery solid walls. In the following section, thermally induced osmotic transport in a confined space will be induced.

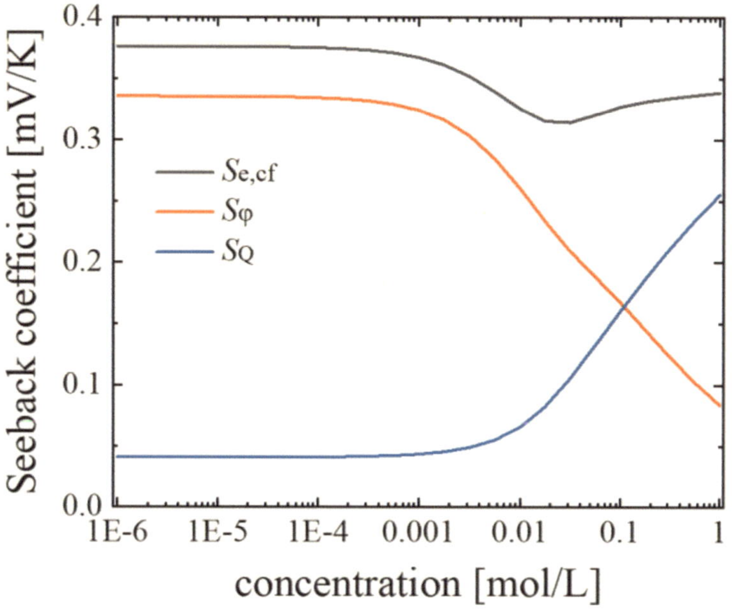


**Figure 2.**
*Dependence of equivalent Seeback coefficient on electrolyte concentration in a 2D slit. The slit height,* H = 20 *nm, and the zeta potential,* ζ = 100 *mV.*

### 2.3 Thermo-osmotic flow in confined space

The confinement of liquid in nanochannels alters its specific enthalpy. For an isothermal system, the liquid flow in confined nanochannels driven by a pressure gradient ($\nabla p$) would produce "heat flux", leading to thermal gradient establishment through the channel. On the other hand, a thermal gradient ($\nabla T$) applied through the nanochannels would also induce the liquid flow, which is the so-called thermos-osmosis. Such a coupling phenomenon is described by the Onsager's linear nonequilibrium thermodynamics [6, 8],

$$\begin{bmatrix} u_{\mathrm{s}} \\ f_{\mathrm{h}} \end{bmatrix} = \begin{bmatrix} \beta_{11} & \beta_{12} \\ \beta_{21} & \beta_{22} \end{bmatrix} \begin{bmatrix} -\nabla p \\ -\dfrac{\nabla T}{T} \end{bmatrix} \tag{20}$$

where $u_{\mathrm{s}}$ is hydrodynamic velocity, $f_{\mathrm{h}}$ is heat transfer flux, and $\beta_{ij}$ $(i, j = 1, 2)$ is the phenomenological coefficient. Here, $\beta_{11}$ characterizes the isothermal flow driven by a pressure gradient, $\beta_{22}$ denotes heat conduction at $\nabla p = 0$, and, according to the Onsager's reciprocity relations, $\beta_{12} = \beta_{21}$, which represents the thermos-osmosis coefficient. The following is intended to obtain the expression for $\beta_{12}$.

Consider a long and thin 2D slit as illustrated in **Figure 1**. To account for the effect of interfacial hydrodynamics, we assume the Navier's slip boundary condition at the slit wall surface [13],

$$u(z_{\mathrm{s}}) = b\left.\frac{\mathrm{d}u}{\mathrm{d}z}\right|_{z=z_{\mathrm{s}}} \tag{21}$$

where $u$ is velocity, $z_s$ is shear plane position, and $b$ is slip length. The slit confinement modifies the specific enthalpy in the liquid, resulting in a thermodynamic driven force, $\delta h(z)$ $(\nabla T/T)$, where $\delta h(z)$ is local excess specific enthalpy. Then, the thermos-osmotic flow can be solved through the Stokes equation,

$$-\eta \frac{d^2 u(z)}{dz^2} = \delta h(z) \frac{\nabla T}{T} \tag{22}$$

where $\eta$ is liquid viscosity and assumed homogenous. Integrating Eq. (22) twice and using boundary conditions, Eq. (21) and $(du/dz)|_{z=0} = 0$, give the thermos-osmotic slip velocity ($u_s$),

$$u_s = \frac{1}{\eta} \int_{z_s}^{0} dz(z - z_s + b)\delta h(z) \frac{\nabla T}{T} \tag{23}$$

Then, according to Eq. (20), the corresponding thermos-osmotic coefficient ($\beta_{12} = \beta_{21}$) is obtained as

$$\beta_{12} = \beta_{21} = -\frac{1}{\eta} \int_{z_s}^{0} dz(z - z_s + b)\delta h(z) \tag{24}$$

When it is positive, liquid flows toward the cold side, and when negative, liquid flows toward the hot side. For $b = 0$, Eq. (24) retrieves the coefficient for the non-slip boundary condition ($\beta_{12}^{no-slip} = \beta_{21}^{no-slip}$). The presence of a slippery boundary magnifies the thermos-osmotic response by [8]

$$\beta_{12} = \beta_{12}^{no-slip}(1 + b/\lambda) \tag{25}$$

where

$$\lambda = \frac{\int_{z_s}^{0} dz(z - z_s)\delta h(z)}{\int_{z_s}^{0} dz\ \delta h(z)} \tag{26}$$

characterizes the length scale of the interfacial liquid layer thickness where the liquid enthalpy is altered by the solid wall.

The thickness $\lambda$ is typically in the order of a few molecular sizes. Thus, it can significantly enhance the thermos-osmotic coupling coefficient on surfaces with ultralow liquid friction, such as on graphitic surfaces, where the slip length reaches over tens of nanometers. Experiments on bare glass and Pluronic coated substrates have revealed thermos-osmotic coefficients ranging from $\beta_{12} \sim 10^{-10}$ m$^2$/s to $10^{-9}$ m$^2$/s [14]. Molecular dynamic simulations have shown that the thermos-osmotic coefficient can be increased by orders of magnitude if the liquid-solid interfacial friction is ultra-low [8]. This hints that the thermos-osmotic flow in slits with partially slippery walls needs to be taken into consideration in the analysis of thermoelectric coupling, which is discussed below.

### 2.4 Thermo-electric response with hydrodynamic slip

The thermos-osmotic coupling analysis above shows that a temperature gradient through confined channels could induce significant liquid flow, especially if the wall has ultra-low liquid friction. Thus, the liquid advection inside those channels needs to be considered in analyzing the thermo-electric response in contrast to Section 2.2. The presence of convection flow modifies the ion flux ($J$) through a confined channel to [10]

$$J_{\pm} = c_{\pm}u - D_{\pm}\left(\nabla c_{\pm} + c_{\pm}S_{\mathrm{T},\pm}\nabla T - c_{\pm}\frac{v_{\pm}e}{k_{\mathrm{B}}T}E_{\mathrm{T}}\right) \tag{27}$$

Assume the double layers inside the 2D slit (see **Figure 1**) are not overlapped. The vanishing ion flux in the $z$ direction also leads to ion concentration distribution following Eq. (11).

To obtain the thermo-electric coupling coefficient, we refer to the Onsager's linear nonequilibrium thermodynamics, which describes the thermo-electric coupling phenomenon by [6, 10]

$$\begin{bmatrix} j_{\mathrm{e}} \\ f_{\mathrm{h}} \end{bmatrix} = \begin{bmatrix} \alpha_{11} & \alpha_{12} \\ \alpha_{21} & \alpha_{22} \end{bmatrix} \begin{bmatrix} -\nabla V \\ -\dfrac{\nabla T}{T} \end{bmatrix} \tag{28}$$

where $j_{\mathrm{e}}$ is electric current density, $-\nabla V$ is the gradient of electric potential, $\alpha_{11}(=\sigma)$ is electrical conductivity, $\alpha_{22}/T(=\kappa)$ is thermal conductivity, and $\alpha_{12} = \alpha_{21} = \alpha_{\mathrm{TE}}$ is the thermoelectric coefficient based on the Onsager's reciprocal relation. Thus, the coefficient $\alpha_{\mathrm{TE}}$ can be obtained by either

$$\alpha_{12} = \frac{-j_{\mathrm{e}}}{\nabla T/T}, \mathrm{at} \nabla V = 0 \tag{29}$$

or

$$\alpha_{21} = \frac{-f_{\mathrm{h}}}{\nabla V}, \mathrm{at} \nabla T = 0 \tag{30}$$

Here, the ion current density ($j_{\mathrm{e}}$) in Eq. (29) is determined according to Eq. (27) by setting $E_{\mathrm{T}} = -\nabla V = 0$,

$$j_{\mathrm{e}} = \frac{ve}{H/2}\int_0^{H/2} (J_{+,x} - J_{-,x})\mathrm{d}z \tag{31}$$

with the concentration profile ($c_{\pm}$) determined by the Poisson-Boltzmann equation and the velocity field ($u$) determined by the thermos-osmotic flow equation [i.e., Eq. (22)].

The heat transfer flux ($f_{\mathrm{h}}$) in Eq. (30) can be derived through [6, 10]

$$f_{\mathrm{h}} = \frac{1}{H}\int_{-H/2}^{H/2} u(z)\delta h(z)\mathrm{d}z. \tag{32}$$

where the velocity field ($u$) is induced by applying a constant external electric field ($E_e$), and described by the Stokes equation,

$$-\eta \frac{d^2u(z)}{dz^2} = c_e E_e \tag{33}$$

with $c_e = ve(c_+ - c_-)$ is charge density, and $E_e = -\nabla V$ is the constant external electric field. Combining the Poisson equation

$$\varepsilon \frac{d^2\varphi(z)}{dz^2} = c_e \tag{34}$$

and boundary conditions, $du(z)/dz|_{z=0} = 0$ and $d\varphi(z)/dz|_{z=0} = 0$, leads to

$$\eta \frac{du(z)}{dz} = \varepsilon E_e \frac{d\varphi(z)}{dz}. \tag{35}$$

At $z = 0$, $\varphi(0) = 0$ and $u(0) = u_{eo}$, which is defined as the electro-osmotic velocity. At $z = H/2$, $\varphi(H/2) = \varphi_0$ and $u(H/2) = b du/dz|_{z=H/2} = b\Sigma E_e/\eta$ with $\Sigma = \varepsilon d\varphi/dz|_{z=H/2}$ as the surface charge density. (Here, the slipping plane position $z_s$ is assumed approximately equal to $H/2$.) Then, the electro-osmotic velocity ($u_{eo}$) is obtained as [10]

$$u_{eo} = \frac{\varepsilon E_e}{\eta} \zeta. \tag{36}$$

where $\zeta$ is effective potential, and

$$\zeta = \varphi_0 + \Sigma b/\varepsilon, \tag{37}$$

indicating the dependence of the $\zeta$ potential on the slippery boundary condition and surface charge density.

With known the velocity profile, the enthalpy density is still required to get the heat transfer flux ($f_h$) according to Eq. (32). According to Ref. [10], the excess enthalpy density ($\delta h(z)$) majorly consists of two parts,

$$\delta h(z) = \delta h_{water}(z) + \delta h_{EDL}(z), \tag{38}$$

where $\delta h_{water}(z)$ is due to the excess enthalpy of water molecular and $\delta h_{EDL}(z)$ is related to the EDL,

$$\delta h_{EDL}(z) = \delta h_{el}(z) + \delta h_{osm}(z) + \delta h_{solv}(z), \tag{39}$$

with $\delta h_{el}(z)$ due to the electrostatic interaction, $\delta h_{osm}(z)$ originating from the osmotic pressure, and $\delta h_{solv}(z)$ coming from the ion solvation.

In general, the excess entropy in confined slits can hardly be directly measured by experiments. Molecule dynamic (MD) simulations have shown that the excess enthalpy of water dominates the overall contributions, which explains the distinct thermoelectric coupling effects as predicted by the Poisson-Boltzmann theory and MD simulations. With Eq. (32), Eq. (36), and Eq. (38), the thermoelectric coupling coefficient ($\alpha_{TE}$) can be obtained according to Eq. (30), and the equivalent Seeback coefficient in slits with hydrodynamic slip boundary conditions, defined by $S_{e,hs} = -\nabla V / \nabla T$ with $j_e = 0$, is expressed by [10]

$$S_{e,hs} = \frac{\alpha_{TE}}{\sigma T} \tag{40}$$

where $\sigma$ is electrical conductivity, consisting of bulk ($\sigma_{bulk}$) and surface conductivity ($\sigma_{surf}$), of which the latter is dominated if $\kappa H << 1$. By scaling analysis of Eq. (30), it is found for surface conductivity-dominated channels that [10]

$$S_{e,hs} \approx \frac{-\delta h \times (2\lambda/H) \times (\varepsilon\zeta/\eta)}{\sigma_{surf}}. \tag{41}$$

Here, λ is described by Eq. (26), indicating the solid-liquid interaction layer thickness. The ζ potential depends on the surface charge density and boundary slip length [see Eq. (37)]. The apparently direct correlation between the equivalent Seeback coefficient and ζ potential implies simultaneously high surface charge density and boundary slip length benefit the enhancement of the thermoelectric coupling by orders of magnitude, as demonstrated by molecular dynamic simulations [10]. It should be remarked that surface charge distribution on highly slippery solid surfaces must be homogeneous. Otherwise, the heterogeneous distribution of those surface charges could deteriorate the slip effect and, consequently, the thermoelectric coupling.

In summary, hydrodynamic slip enhances the thermo-osmotic coupling, inducing significant convection flow inside confined slits. This contributes to the ion fluxes driven by a temperature gradient, or, in other words, the thermoelectric coupling effect. The thermoelectric coefficient can be obtained according to the Onsager's linear nonequilibrium thermodynamics. The scaling analysis of the equivalent Seeback coefficient shows the thermoelectric coupling effect that can be enhanced by maximizing the surface charge density and slip length simultaneously. In the next, the current experimental achievements and challenges in nanofluidic thermoelectricity will be discussed.

## 3. Experimental achievements, challenges, and prospective

### 3.1 Thermoelectric performance of nanofluidic devices

Although considerable efforts have been made to improve the thermoelectric performance of semiconductors, ionic gels, etc., experimental investigation of the thermoelectric coupling of nanofluidic devices still lies in its fetal stage. A typical experimental setup for nanofluidic thermoelectricity is schematically shown in **Figure 3**. A membrane with nanofluidic channels is separated between two liquid reservoirs filled with aqueous electrolyte. A temperature gradient is applied across the nanofluidic membrane to produce thermoelectricity. To improve the thermal energy conversion efficiency, the key lies in optimizing the surface chemistry and geometric structure of nanofluidic channels.

A previous study of electrokinetic transport through ion rectifying channels provides inspiration for the design of thermoelectric nanofluidics. The electricity generation due to electrolyte transport through nanochannels driven either by thermal gradient or by pressure gradient requires the efficient separation of positive and negative ions. Thus, nanochannels with ion rectifying properties are expected to be able to work as well-

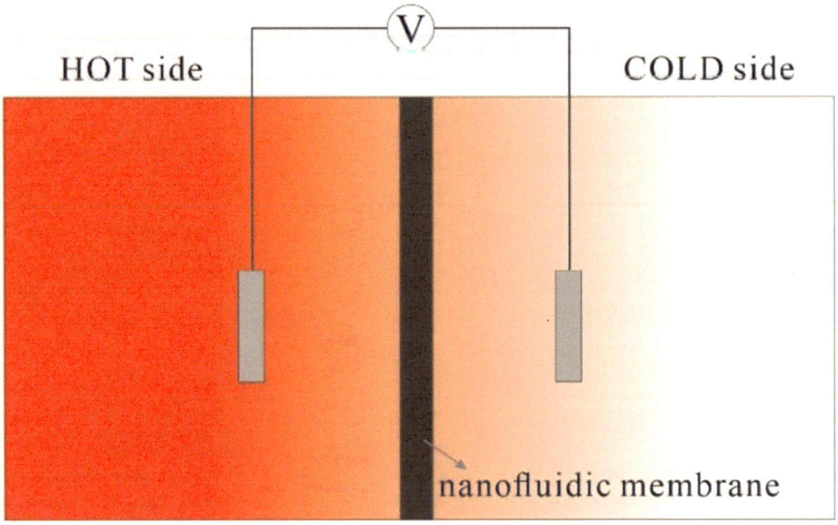


**Figure 3.**
*Schematics of experimental setup for thermoelectricity measurements through nanofluidic membranes.*

performed thermal energy converters. Attempts have been made with chemically and/or geometrically asymmetric nanochannels to probe the thermoelectric coupling effect [15–17]. By partially coating silicon dioxide nanochannels with hydrophobic molecular, linear dependence of the thermoelectric current on the temperature difference is observed, which is attributed to the slippage-induced thermo-osmotic transport [15]. Asymmetric cone-shaped silica nanofluidic channels with dopamine-grafted inner surface have also been shown to be able to generate thermoelectricity with power throughput reaching 25.48 pW per channel at a temperature difference of 40°C [16]. However, the average equivalent Seeback coefficient with a value around 0.4 mV/K is still pretty low as compared with conventional thermoelectric semiconductors.

Smart biological system has the ability of thermosensation relying on ion channels on cell membranes [18], motivating the bionic design of highly temperature-sensitive nanofluidic membranes. Chen et al. [19] constructed a permselective ionic membrane by stacking ultrasmall silica nanochannels of around 2.3 nm in diameter on track-etched poly(ethylene terephthalate) conical nanochannels of around 10–15 nm in the small side. The 2.3-nm silica channels with negatively charged surface preferentially allow the transport of positive ions. A temperature sensitivity of around 0.7 mV/K is demonstrated using such hybrid nanochannel membranes. Ionic covalent organic framework (COF) with pore size below 1 nm, close to that of biological ion channels, has also attracted great attention for temperature sensation application [20]. The high charge density inside the sub-nanometer COF pores enables enhanced thermoelectric response with equivalent Seeback coefficient reaching around 1.27 mV/K. Although this sensitivity is relatively larger than that in ultrasmall silica nanochannels, it is still significantly weaker than those of common thermoelectric semiconductors and ionic gels.

### 3.2 Main challenge and prospective

Theoretical analysis of thermoelectric response in 2D slits with partial slip boundary conditions has shown that simultaneously high surface charge density and slip length can improve the thermoelectric coupling effect to a level comparable to that of

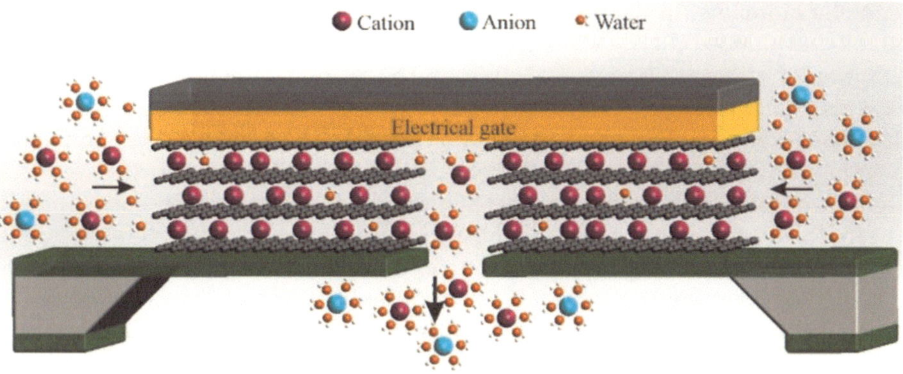


**Figure 4.**
*Schematics of ultrafast ionic transport through graphene channels under electrostatic gating [23].*

common thermoelectric semiconductors. However, current achievements of thermal sensitivity of nanofluidics channels are still quite low, generally in the order of 1 mV/K or even below. The main challenge lies in the difficulty to realize the ultrahigh slip effect on highly charged solid surface. In the inner surface of silica and COF nanochannels, the charge density is usually quite high, but the slip length is almost vanishing or even negative. That is because those charges are heterogeneously distributed on the channel surface, which reduces the slippery effect. Molecular dynamic simulation reveals that the slip boundary condition remains almost unchanged if the surface charges are homogenously distributed [10, 21]. This paves a way to achieve both highly slippery and highly charged nanochannels for enhanced thermoelectric conversion.

Ultrafast fluidic and ionic transport on highly charged graphitic surfaces, e.g., carbon nanotubes [22] and graphene [23], have recently been observed experimentally. In contrast to solid surfaces like silica, graphitic surfaces are usually atomically smooth and exhibit ultralow friction to the water. Although graphitic surfaces are inert to chemical modification, external electrical gating can tune the surface charge density, which shows almost no effect on the slippery property of the surface. Thus, electrostatic gating in atomic-scale graphene channels enables ultrafast and tunable ionic transport with an effective diffusion coefficient reaching two orders of magnitude higher than in bulk water [23]. Moreover, the self-assembly of 2D material flakes easily enables large-scale membrane fabrication. Therefore, highly improved thermoelectric coupling efficiency can be expected in atomically smooth 2D material channels with enhanced charge density (**Figure 4**).

## 4. Conclusions

Emergent and efficient thermoelectric conversion techniques are highly required to harvest low-grade heat in aqueous solution. Nanofluidic systems show great potential to be used for effective thermoelectric energy conversion, which, however, has been poorly explored so far. In bulk electrolyte, the Soret-type thermophoretic motion of positive and negative ions with different ionic heat of transport establishes an electrical field under a thermal gradient. However, the equivalent Seeback coefficient is generally quite low, in the order of 0.1 mV/K. In confined nanochannels, the confinement induced thermoelectricity is dominated over the Soret effect with the assumption of no convection flow. However, the overall equivalent Seeback coefficient is also quite low.

It is remarked that, in nanochannels with partial slip boundary conditions, the convection flow of liquid under a temperature gradient, that is, thermo-osmosis, can be significant and needs to be considered in the thermoelectric conversion analysis. It has been demonstrated by molecular dynamic simulations that simultaneously high slip length and surface charge density contribute to orders of magnitude enhanced thermoelectric coupling coefficient. However, the realization of such a kind of nanochannel surface is quite challenging. It is the main reason why temperature sensitivity of only around 1 mV/K is achieved even in sub-nanometer channels with ultrahigh surface charge density.

It is revealed that heterogeneously distributed charges deteriorate the slippery boundary condition. Thus, homogeneous distribution of surface charges is required to achieve constant low liquid friction. The observation of ultrafast ionic transport in graphene channels under electrostatic gating points out a new direction to combine atomically smooth 2D materials and the electrostatic gating technique for efficient thermoelectric energy conversion in the future.

## Acknowledgements

This work is supported by the National Program on Key Basic Research Project of China (Grant No. 2022YFA1203400), the National Natural Science Foundation of China (Grant No. 12272159), and the National Natural Science Foundation of Guangdong Province (Grant No. 2023A1515012592).

## Conflict of interest

The authors declare no conflict of interest.

## Author details

Yahui Xue
Center for Complex Flows and Soft Matter Research and Department of Mechanics and Aerospace Engineering, Southern University of Science and Technology, Shenzhen, China

*Address all correspondence to: xueyh@sustech.edu.cn

## References

[1] Hamid Elsheikh M, Shnawah DA, Sabri MFM, Said SBM, Haji Hassan M, Ali Bashir MB, et al. A review on thermoelectric renewable energy: principle parameters that affect their performance. Renewable and Sustainable Energy Reviews. 2014;**30**: 337-355. DOI: 10.1016/j.rser.2013.10.027

[2] Kishore RA, Priya S. A review on low-grade thermal energy harvesting: materials, methods, and devices. Materials (Basel). 2018;**11**:1433. DOI: 10.3390/ma11081433

[3] Schoch R, Han J, Renaud P. Transport phenomena in nanofluidics. Reviews of Modern Physics. 2008;**80**:839-883. DOI: 10.1103/RevModPhys.80.839

[4] Bocquet L, Charlaix E. Nanofluidics, from bulk to interfaces. Chemical Society Reviews. 2010;**39**:1073-1095. DOI: 10.1039/b909366b

[5] Derjaguin B, Sidorenkov GP. On thermo-osmosis of liquid in porous glass. Comptes Rendus del'Academie des Sciences de'lURSS. 1941;**32**:622-626

[6] Derjaguin BV, Churaev NV, Muller VM. Surface Forces. Boston: Springer; 1987. p. 369. DOI: 10.1007/ 978-1-4757-6639-4

[7] Bregulla AP, Würger A, Günther K, Mertig M, Cichos F. Thermo-osmotic flow in thin films. Physical Review Letters. 2016;**116**:188303. DOI: 10.1103/ PhysRevLett.116.188303

[8] Fu L, Merabia S, Joly L. What controls thermo-osmosis? molecular simulations show the critical role of interfacial hydrodynamics. Physical Review Letters. 2017;**119**:214501. DOI: 10.1103/ PhysRevLett.119.214501

[9] Würger A. Thermal non-equilibrium transport in colloids. Reports on Progress in Physics. 2010;**73**:126601. DOI: 10.1088/0034-4885/73/12/126601

[10] Fu L, Joly L, Merabia S. Giant thermoelectric response of nanofluidic systems driven by water excess enthalpy. Physical Review Letters. 2019;**123**: 138001. DOI: 10.1103/ PhysRevLett.123.138001

[11] Dietzel M, Hardt S. Thermoelectricity in confined liquid electrolytes. Physical Review Letters. 2016;**116**:225901. DOI: 10.1103/ PhysRevLett.116.225901

[12] Agar JN, Mou CY, Lin JL. Single-ion heat of transport in electrolyte solutions: a hydrodynamic theory. The Journal of Physical Chemistry. 1989;**93**:2079-2082. DOI: 10.1021/j100342a073

[13] Lauga E, Brenner M, Stone H. Microfluidics: the no-slip boundary condition. In: Springer Handbook of Experimental Fluid Mechanics. Berlin, Heidelberg: Springer; 2007. p. 1219. DOI: 10.1007/978-3-540-30299-5_19

[14] Bregulla AP, Würger A, Günther K, Mertig M, Cichos F. Thermo-osmotic flow in thin films. Physical Review Letters. 2016;**116**:188303. DOI: 10.1103/ PhysRevLett.116.188303

[15] Li L, Wang Q. Thermoelectricity in heterogeneous nanofluidic channels. Small. 2018;**14**:1800369. DOI: 10.1002/ smll.201800369

[16] Zhao XL, Li L, Xie WY, Qian YC, Chen WP, Niu B, et al. pH-regulated thermo-driven nanofluidics for nanoconfined mass transport and energy conversion. Nanoscale Advances. 2020; **2**:4070-4076. DOI: 10.1039/d0na00429d

[17] Wang R, Sun Y, Zhang F, Song M, Tian D, Li H. Temperature-sensitive artificial channels through pillar[5] arene-based host–guest interactions. Angewandte Chemie International Edition. 2017;**56**:5294-5298. DOI: 10.1002/anie.201702175

[18] Buijs TJ, McNaughton PA. The role of cold-sensitive ion channels in peripheral thermosensation. Frontiers in Cellular Neuroscience. 2020;**14**:262. DOI: 10.3389/fncel.2020.00262

[19] Chen K, Yao L, Su B. Bionic thermoelectric response with nanochannels. Journal of the American Chemical Society. 2019;**141**:8608-8615. DOI: 10.1021/jacs.9b03569

[20] Zhang P, Chen S, Zhu C, Hou L, Xian W, Zuo X, et al. Covalent organic framework nanofluidic membrane as a platform for highly sensitive bionic thermosensation. Nature Communications. 2021;**12**:1844. DOI: 10.1038/s41467-021-22141-z

[21] Xie Y, Fu L, Niehaus T, Joly L. Liquid-solid slip on charged walls: the dramatic impact of charge distribution. Physical Review Letters. 2020;**125**: 014501. DOI: 10.1103/ PhysRevLett.125.014501

[22] Xue Y, Yang Y, Sun H, Li X, Wu S, Cao A, et al. A switchable and compressible carbon nanotube sponge electrocapillary imbiber. Advanced Materials. 2015;**27**:7241-7246. DOI: 10.1002/adma.201502837

[23] Xue Y, Xia Y, Yang S, Alsaid Y, Fong KY, Wang Y, et al. Atomic-scale ion transistor with ultrahigh diffusivity. Science. 2021;**372**:501-503. DOI: 10.1126/ science.abb5144

Chapter 7

# Modern Physics of the Thermoelectric Phenomena: Achievements and Problems

*Gulmurza Abdurakhmanov, Dibya Prakash Rai and Gulbahor Vokhidova*

## Abstract

This chapter discusses internal discrepancies of contemporary conceptions of physics of thermoelectric phenomena (Seebeck, Peltier, and Thomson effects). These conceptions contradict also with experimental data obtained in a wide range of temperature for various materials (pure metals, alloys, Si, Ge, intermetallic and oxide compounds, borides, and silicides). One of these contradictions arises from the energy conservation law and definition of the Seebeck coefficient—the last cannot exceed 86.25 μV/K in any material. This limitation is met in metals and alloys, while in nonmetallic materials it exceeded hundreds and thousands of times. Experimental temperature dependence of the Seebeck coefficient demonstrates the polarity reversal and sharp extrema (increases up to 100–1000 times) for various materials, which are not followed from theory. Constancy of the Seebeck and Peltier coefficients (underlying the definitions of thermoEMF and Peltier heat) contradicts with Thomson formulae requiring temperature dependence of these coefficients (otherwise the Thomson effect is absent in any materials). The role of structural (spatial) inhomogeneity of the thermoelectric material and the wave nature of thermal radiation are discussed for potential physical mechanism of thermoEMF generation. Extension of expressions for charge and thermal energy flow to take into account nonlinear properties leads to huge mathematical complications.

**Keywords:** Seebeck effect, Peltier effect, Thomson effect, thermopower, thermoelectric figure of merit, phonon drag, sign inverse, irreversible thermodynamics, Boltzmann equation, Onsager's relations, structure transitions, high-frequency electromagnetic waves, inhomogeneity

## 1. Introduction

The thermoelectric materials are the new energy efficient materials for the green energy harvest. Owing to (i) threat to exhaustion of fossil fuels (oil, gas, and coal), so there is a need to create efficient and cheap methods of using renewable energy sources and (ii) the expansion of technical and technological capabilities of mankind, which leads to an increase in the amount of waste heat and pollution of the environment. This worsens the conditions of human existence and forces to look for methods of using waste heat; (iii) creation of the theory of "phonon glass, electronic crystal" [1] and the discovery of a new class of thermoelectric materials, the oxide compound $NaCo_2O_4$ having high value of thermopower [2], the research interest on the thermoelectric phenomena in various materials and structures has increased dramatically in the last two decades.

Scientific aspect of this interest is also important because thermoelectricity:

i. allows us to have information on the state of the electronic subsystem in various materials and structures;

ii. has served an experimental basis for verification of concepts of equilibrium and nonequilibrium thermodynamics;

iii. can serve as a scientific basis for creating new thermoelectric materials used in devices to convert waste heat and solar radiation into electricity.

Over two centuries of research on thermoelectric phenomena, many famous scientists have contributed to their study and practical applications. As a result, a physical concept of these phenomena was developed, which made it possible to explain the temperature dependence of the coefficient of thermoEMF (Seebeck coefficient) $S$ in metals and semiconductors at high temperatures $T$, where the dependence $S(T)$ is weak. These concepts are based on the laws of linear nonequilibrium thermodynamics and the solution of the Boltzmann kinetic equation (mainly with respect to the relaxation time approximation $\tau$). However, such nonlinear properties of $S(T)$ such as changes in sign and extreme value at low temperatures are often considered as anomalies not worthy of attention and therefore not reflected in these concepts.

A close examination of theoretical conceptions, including phonon drag, and a comparison of their results with experiment reveals a number of internal disagreements and inconsistencies. For example, researchers have contrary definitions of "absolute thermopower" and "differential thermopower" [3, 4]. The present chapter will discuss (1) the main theoretical and experimental results of thermoelectric phenomena, (2) the internal inconsistencies of the theoretical concepts and their discrepancies with the experiment, (3) the possible causes of these inconsistencies and discrepancies, and (4) the possible ways to overcome them. The effects of magnetic field as well as anisotropy of material will not be considered.

## 2. Theoretical basics

There are three thermoelectric phenomena (**Figure 1**):

1. Seebeck effect – generation of thermoEMF $\Delta\varphi$ in material due to the temperature difference $\Delta T = T_H - T_C$ (discovered in 1821).

2. Peltier phenomenon – cooling ($Q_P < 0$) or heating ($Q_P > 0$) of contact of different conductors, connected in series, according to the direction of electrical current $I$ (discovered in 1834).

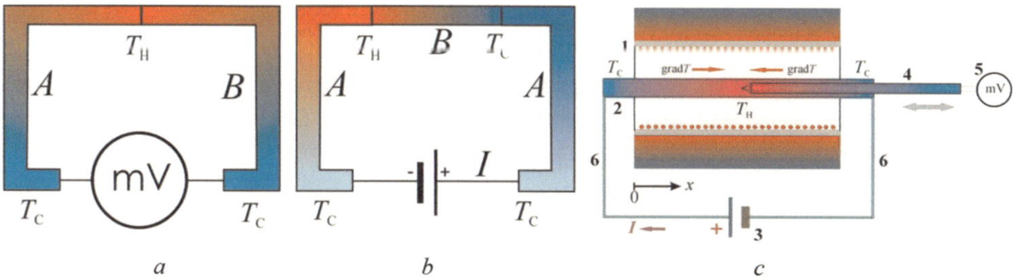


**Figure 1.**
*Setup for observing Seebeck (a), Peltier (b), and Thomson (c) effects.*

3. Thomson phenomenon (1854) – evolution ($Q_T > 0$) or absorption ($Q_T < 0$) of additional heat in homogenous conductor, which is simultaneously affected by a temperature difference $\Delta T$ and an electric current $I$. Here, $T_H$ and $T_C$ are temperatures of the hot and cold ends of the sample, respectively. The most studied and frequently used phenomena are the Seebeck phenomenon (generators of electrical energy and thermocouples to measure temperature in wide range as well as thermopiles to measure thermal radiation intensity) and the Peltier phenomenon (refrigerators, heaters, and heat pumps), which are considered mutually reversible, while the Thomson phenomenon is of scientific interest mainly because of its small value.

The thermoelectric properties of a material characterized by Seebeck $S$ (thermoEMF), Peltier $\pi$, and Thomson $\tau$ coefficients:

$$\Delta\varphi = S\Delta T, \quad (1)$$

$$Q_P = \pi I, \quad (2)$$

$$Q_T = \tau I \Delta T. \quad (3)$$

The additional heat quantities $Q_P$ and $Q_T$ in expressions (2) and (3) are linear functions of current $I$, as opposed to Joule heat $Q_J = I^2R$ of the sample of resistance $R$. Therefore, the emanation or absorption of $Q_P$ and $Q_T$ depends on the direction of the current. Thomson effect is considered to be positive ($Q_T > 0$) and causes heating of the conductor if the direction of the current ($I$) coincides with the gradient of the temperature ($T$). Samoylovich and Korenblit [3, 5] reported that the emanation of the heat is proportional to a current, which is caused by the inhomogeneity of the system through which the current flows. This inhomogeneity cannot be of chemical nature only but also due to difference in their structure. For example, thermoEMF is generated at the interface of mechanically deformed and non-deformed (hardened and non-hardened, quenched and unquenched) sections of the homogeneous conductor [6] as well.

The definitions of the Seebeck, Peltier, and Thomson effects, written in the form (1)–(3), imply by default that coefficients $S$ and $\tau$ do not depend on $T$ (or $\Delta T$) while $\pi$ and $\tau$ do not depend on $I$.

Thomson's thermodynamic theory [7] has established an intercorrelation of coefficients $S$, $\pi$, and $\tau$ (the first and the second thermoelectric relations, respectively):

$$\tau = T(dS/dT), \quad (4)$$

$$\pi = TS. \tag{5}$$

Further development of thermodynamics of the thermoelectricity has led to reciprocity relations [8, 9] of kinetic coefficients $L_{12} = L_{21}$ in the generalized linear equations for charge flow $j(T, E)$ and heat flow $q(T, E)$:

$$j = -L_{11}\nabla\frac{\mu}{T} + L_{12}\nabla\frac{1}{T}, \tag{6}$$

$$q = -L_{21}\nabla\frac{\mu}{T} + L_{22}\nabla\frac{1}{T} - \mu j, \tag{7}$$

where $\mu = \mu_0 + e\varphi$ is the electrochemical potential consisting of chemical $\mu_0$ and electric $\varphi$ parts.

The linear Eqs. (6) and (7) for the charge and heat flows, bringing to Onsager's reciprocity relations [8, 9], deny the possibility of thermoEMF generation in the homogeneous isotropic conductor (Magnus's law, 1851; see also [3, 5]). Thus, Samoylovich and Korenblit [3, 5] have emphasized that the assumption on reversibility of thermoelectric heat (Peltye and Thomson) does not follow from the second principle of thermodynamics and is the additional assumption. On the other hand, for developing Eqs. (4) and (5), W. Thomson considered the thermoelectric phenomena as reversible that is not rather reasonable. This idea of Thomson's theory was criticized by Boltzmann as it has fail to confirm the results of various experiments [3, 5]. We will see below that the experimental data used for confirmation, which almost do not contain any nonlinearity. Herein, the nonlinearities (see, for example, [6]) were excluded intentionally as anomalies.

The situation is that the researchers ignore irreversibility of thermoelectric effects [3]: "Thomson applied the first and second principles of thermodynamics to the analysis of thermoelectric phenomena, considering thermoelectric processes to be reversible. The proportionality of the Peltier and Thomson heats to the strength of the electric current and, consequently, the fact that when the direction of the current changes, instead of releasing heat, there will be absorption, or, conversely, makes natural the assumption that in the case of the Thomson and Peltier effects, we are dealing with processes that are reversible in the thermodynamic sense of the word." It can be seen here that the reversibility of thermoelectric phenomena is provided by excluding from consideration the fact that when an electric current passes through a homogeneous conductor placed in a medium with a uniform temperature, Joule heat $Q_J = I^2R$ is emanated and dissipated in the surrounding medium. Joule heat is the main part of the energy released in the conductor. One can say that the emanation of Joule heat when an electric current flows through a homogeneous conductor placed in a medium homogeneous in temperature is common, while no one has seen the reverse process - the generation of an electric current in a homogeneous conductor due to the uniform temperature along the conductor.

It has to be noted that in definition of Seebeck coefficient (1) $e\Delta\varphi$ represents (**Figure 1**) excess Coulomb energy of electrons on the cold end of a sample relative to the hot end, arising from their diffusion due to surplus of thermal energy $k_B\Delta T$ on the hot end relative to the cold one ($e$ is the electron's charge). Therefore, it is reasonable to write $e\Delta\varphi \le k_B\Delta T$, which in a steady state without a current in an external circuit (an open-circuit mode) leads to the maximum value of Seebeck coefficient: $|S| \le k_B/e = 86.25\ \mu V/K$. Here $k_B$ is Boltzmann's constant. It is important to note that this limit does not depend on both the studied substance and on the temperature range.

Apparent expressions for $S(T)$ of metals and semiconductors have derived from Boltzmann kinetic equation [10]:

For metals, (degenerate electron gas)

$$S(T) = -\frac{\pi^2}{3}\frac{k_B}{e}\frac{k_B T}{\mu}\left(r + \frac{3}{2}\right), \tag{8}$$

For semiconductors (nondegenerate electron gas)

$$S(T) = -\frac{k_B}{e}\left[\frac{\mu}{k_B T} - \left(r + \frac{5}{2}\right)\right]. \tag{9}$$

Here, $r$ is the exponent of momentum relaxation. These expressions are applicable for higher temperatures (region I in **Figure 2**), while at lower temperatures, there are polarity reversal and extremum(s) of $S(T)$. Increasing of $|S(T)|$ at low temperatures (region II in **Figure 2**) is usually explained by phonon drag [10–13] (Herring model, see also Seeger [4]):

$$S_{ph} = \frac{k_B}{e}\left(\frac{m^* s^2}{k_B T}\right)\frac{\langle \tau_e \tau_{ph}/\tau_{eph}\rangle}{\langle \tau_e \rangle} \propto T^{-7/2}. \tag{10}$$

Here, $s$ is the sound velocity, $\tau_e$ – full relaxation time of electrons momentum, $\tau_{eph}$ – relaxation time of electrons momentum due to scattering on phonons, and $\tau_{ph}$ – relaxation time of phonons averaged over momentum. For moderate temperatures, when phonon-electron scattering obeys Herring model, $S_{ph} \sim T^{-7/2}$ [14]. Gurevich and Mashkevich [15] showed that phonon drag effects not only $S$ but also the electrical conductivity of semiconductors.

This mechanism implies that phonons (thermal vibrations of atoms) move from the hot end of the sample to the opposite end simultaneously with electrons. Phonons

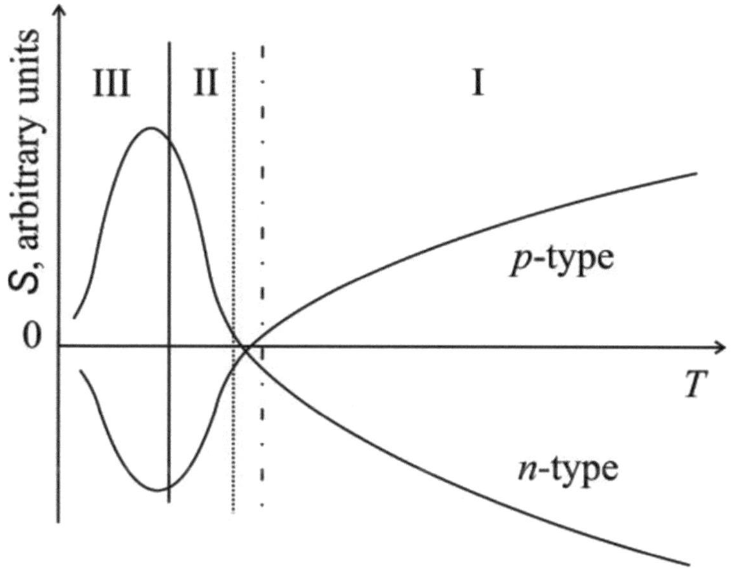


**Figure 2.**
*Typical temperature dependence of the thermopower* S(T) *(schematic).*

average velocity ($\approx 5 \cdot 10^3$ m/s) in solids is essentially higher than electrons average one ($\approx$ 40 m/s for diffusion in Si). Therefore, phonons, colliding with electrons, pass them part of impulse, creating an additional charge flow and corresponding thermoEMF. The estimation of this additional thermoEMF gives 400 mV/K up to 10 V/K [4].

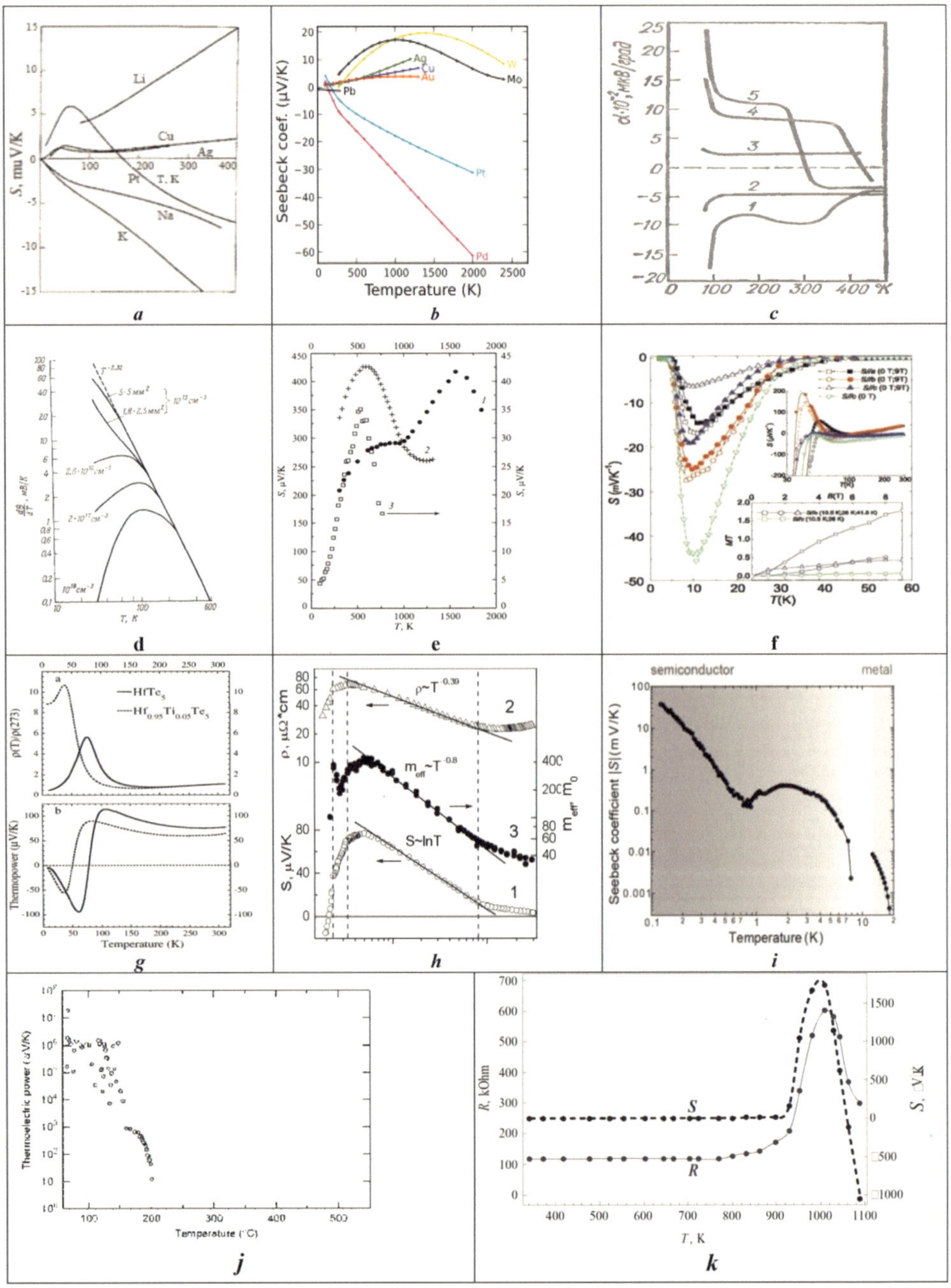


**Figure 3.**
*Temperature dependence of thermopower in: Metals at low (a) [16] and high temperatures (b) [17]; (c) germanium [18]; (d) silicon [19]; (e) borides $AlB_{12}$ (1), $B_{14}Si$ (2), FeB29:5 (3) [20]; (f) $FeSb_2$ [21]; (g) $HfTe_5$ and $Hf_{0.95}Ti_{0.05}Te_5$ monocrystals [22]; (h) $CeB_6$ thermopower S(T) (1), specific resistance ρ(T) (2), effective mass m*(T) (3) [11]; (i) $(TMTSF)_2PF_6$ [23]; (j) $TlInSe_2$ [24]; (k) the doped silicate glass [25].*

Based on thermodynamic considerations, it is assumed that near 0 K (region III in **Figure 2**), the Seebeck coefficient tends to zero. This assumption is experimentally confirmed in many pure metals and alloys and nonmetallic compounds (**Figure 3a, b, d-g**), but a deviation from it is observed in some doped alloys and nonmetallic materials (**Figure 3c, h, i,** and **4**).

The opinions of researchers are differing about where the thermoEMF occurs. In many reports [4, 27], it is believe that thermoEMF occurs in the volume of a homogeneous material due to a temperature drop (absolute thermopower). Others [3, 5] associate the appearance of thermoEMF with the inhomogeneity of the material. Ioffe [28] related part of $S$ to the contacts. However, this question is still open due to the lack of direct experimental evidence of one or the other. From an experimental point of view, absolute thermopower is irrelevant since it cannot be probed directly from a voltmeter.

The relationship of $S$ with the electron energy structure (energy levels) of a metal is given by the Mott formula [29]:

$$S = \frac{\pi^2}{3}\frac{k_B^2 T}{e}\left[\frac{d\ln\sigma(E)}{dE}\right]_{E\ddagger\mu}, \tag{11}$$

$$\sigma(E) = -\frac{2e^2}{3m^*}\int_0^\infty g(E)\tau_e\frac{df_0(E)}{dE}dE, \tag{12}$$

$g(E)dE = \frac{4\pi(2m^*)^{3/2}}{h^3}dE$ is the density of electronic states; $m^*$ is an effective mass of charge carriers; $h$ – Plank constant, $f_0(E)$ is the Fermi distribution function

$$f_0(E) = \frac{1}{1+e^{(E-\mu)/k_BT}}. \tag{13}$$

Comparing distribution of energy flow in the thermopile branches (**Figure 1**) in the framework of Thomson's theory and the computational results, Spry [30] concludes that Thomson's theory is based on the erroneous precondition. Hence, for the description of the thermoelectric phenomena, four parameters ($\sigma$, $\kappa$, $S$, and $\pi$) of a material suffice, and Thomson's phenomenon can be neglected.

## 3. Experimental results

For almost two-centuries history of a thermoelectricity, the thermoelectric properties of the various materials have been investigated and agreement of experimental data with expressions (1), (8) and (9) at high temperature, assuming weak variation in the values of $S$, $\pi$ and $\tau$ while significant deviations from these expressions have been observed at a low temperatures. Temperature dependence $S(T)$ of various materials are shown in **Figures 3–8**.

Let us note some features of the experimental data presented in **Figures 3–8**: (1) sharp changes of $S(T)$ at low temperatures; (2) polarity reversal of $S$ in many cases; (3) the maximum value of $S$ in many materials essentially exceeds the limit of 86 μV/K. For example, in the compound $TlInSe_2$ $S \approx 10$ V/K at $T \leq 100$ K (**Figure 3j**) and in the organic semiconductor $(TMTSF)_2PF_6$, $S(T)$ increases at $T \to 0$ K (**Figure 3i**) instead of reduction according to the theory. Properties (1) and (2) have been found in metals more than 100 years ago [3, 5]. Consequence of these features signifies a strong

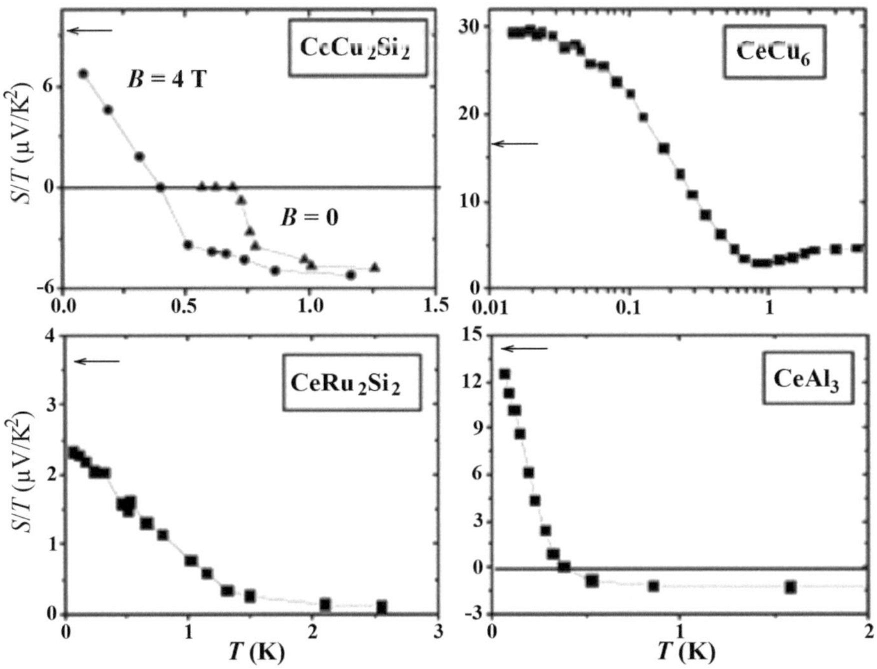


**Figure 4.**
*S/T as a function of temperature for four different Ce-based compounds [26] using previously published data by three different groups. In each panel, the horizontal vector points to the value corresponding to γ/(NAve). Note the semilogarithmic scale in the case of $CeCu_6$.*

temperature dependence of Peltier $\pi$ and Thomson $\tau$ coefficients, defined by relations (4) and (5).

Seeger [4] gives dependence of $S(T)$ in silicon (**Figure 3d**) where $S(T) \approx 100$ mV/K is reached at $T = 30$ K, as confirmation of phonon drag. However, there is a question: what is a source of additional energy of electrons to carry out the energy conservation law? Such high magnitude of $S$ has observed in Ge by Geballa at about 20 K (**Figure 5**) [14].

There are many results on exotic materials as well as on nanomaterials. Li et al. [31] investigated lightly doped diamond (**Figure 6**) and high $S$ value (about 100 mV/K at 100 K) as well as its temperature dependence explained in terms of phonon drag.

The nanoparticle size ($d$) dependence of $S$ was first observed by J. Singh et al. [34] and Soni and Okram (**Figure 7**) [35]. It was found that the magnitude of $S$ increases as $d$ decreases and achieves the large value about 100 mV/K at 20 K for $d \leq 7$ nm, while for bulk nickel, this is about few μV/K. Authors have tried to explain that this huge value of $S$ is a result of the enhanced grain-boundary scattering combined with quantum confinement. It is also noteworthy that the nanoparticles did not show sign reversal, in contrast to bulk nickel.

Nanotechnology approach has raised $ZT$ value up to ~3.6 at 1000 K for well-known Si-Ge solid solution [37], a high-temperature thermoelectric in use. Supersaturated solid solutions of Si-Ge containing ~1 at.% Fe and 10 at.% P are prepared by high-energy ball milling. The bulk samples consisting of ultrafine nano-crystallites (9.7 nm) are obtained by the sophisticated "low-temperature and high-pressure sintering process." Despite that the electrical resistivity is slightly high due to the localization of electrons associated with the highly disordered structure and low

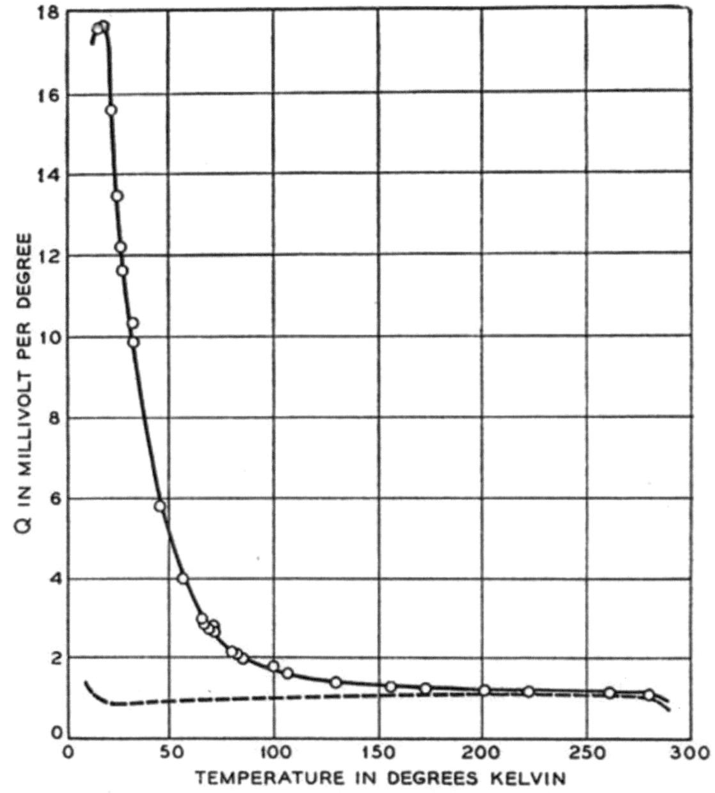


**Figure 5.**
*Thermopower Q (as is in original, instead of* S – G. A.*) of a filament of* p-germanium (Geballe, *unpublished), compared with the Q of Eq. (9) [14]. The room temperature carrier concentration is 5.0 $10^{13}$ $cm^{-3}$. The dashed curve is computed from (9), using the cyclotron masses; the bend over at the right takes account of intrinsic conduction near room temperature by placing the EMF for the holes in parallel with one of similar form for the electrons.*

electrical density of states near the Fermi energy, a very low thermal conductivity $\kappa$ ($< 1$ W $m^{-1}$ $K^{-1}$) and a very large value of Seebeck coefficient $|S|$ exceeding 470 μV $K^{-1}$ have been achieved in association with the nanostructuring and the Fe 3*d* impurity states, respectively.

Hinterleitner et al. [38] have estimated $ZT$ = 5–6 at 370 K for 1 μm thick films of the metastable Heusler alloy $Fe_2V_{0.8}W_{0.2}Al$. Rising of $ZT$ is mainly due to very large slope of density of states near Fermi level. For such materials, $ZT$ = 20 was predicted theoretically [39].

An interesting result has been achieved in $Cu_2Se$ [36] – in the region of structure transition, thermoelectric figure of merit $ZT$ has risen up to 470 – highest value in the history of thermoelectricity. On this basis, one can say that (1) an absence of any fundamental limitations for the value of $ZT$ has been experimentally conformed; (2) the ideal heat engine has been created having efficiency of Carnot's cycle; (3) structure of the thermoelectric is crucial for thermoelectric phenomena. Unfortunately, this heat engine is no practical applicable because of the very narrow (order of few K) working interval of temperature, and low total efficiency.

One possible way to improve thermoelectric properties is using two-dimensional materials with negative correlation between electrical and thermal

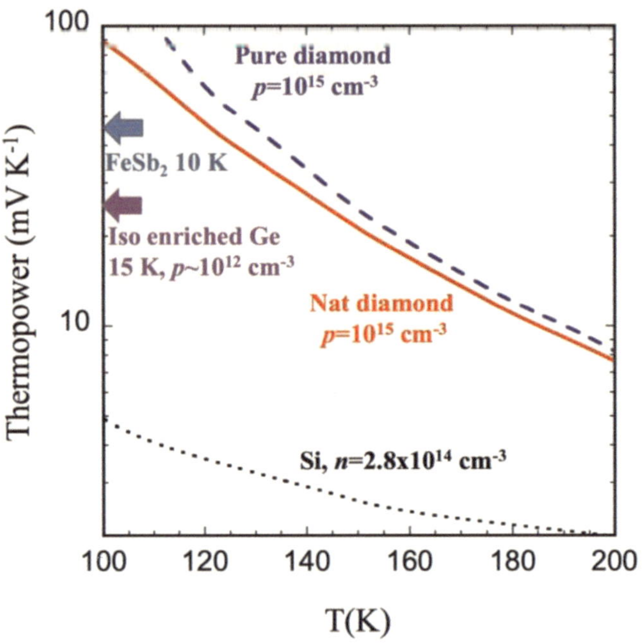


**Figure 6.**
*Temperature dependence of thermopower of diamond with hole density* p = $10^{15}$ *cm*$^{-3}$*, calculated from first principles [31]. Solid red curve is for natural diamond (1.1%* $^{13}C$*). Dashed blue curve is for isotopically pure diamond (100%* $^{12}C$*). Dotted black curve is the calculated magnitude of the thermopower for silicon at electron density 2.8* $10^{14}$ *cm*$^{-3}$ *[32]. Blue arrow shows the maximum measured thermopower for* $FeSb_2$ *at 10 K [21], while purple arrow shows the maximum measured thermopower for isotopically enriched Ge at 15 K [33].*

conductivity [40], which leads to increasing of $\sigma$ while $\kappa$ decreases. For the 16-nm-thick, two-dimensional $SnS_2$ nanosheets obtained at 300 K thermopower 34.7 mV/K and $ZT \approx 0.13$, last value is ~1000 times greater than previously reported bulk single-crystal $SnS_2$.

The silicate glass doped by ruthenium dioxide [25, 41, 42] stands by itself among the materials of thermoelectric properties investigated. This glass has metallic properties ($S \approx 10-20$ μV/K, $\rho \propto T^{\alpha}$, $\alpha = 1-2$) near room temperature, and the maximum of $S(T) \approx 1.7$ mV/K is observed at $T \approx 970-1000$ K instead of at low temperatures (**Figure 3k**, more detailed in **Figure 9**). Such behavior, being accompanied by a similar change of resistance $R(T)$, is assumed to be a consequence of structural transformations in silicate nanocrystals leading to change of energy gap between the impurity band and the top of the valence band of glass [25, 43]. This energy gap is the pseudogap in reality, and nanocrystals act as localization centers for charge carriers [44, 45].

It has been reported in the nineteenth century that $S$ depends not only on the composition of the materials but also on its mechanical deformation (change in structure) [6]. Mechanism of this effect remains unclear. Effect of isotopic substitution of impurities in metals on $S$ remains unexplained for many years [27]. Progress in thermoelectric materials in last two decades (**Figure 10**) based on nanotechnology – nanosize in homogeneities generated in the material to enhance phonon scattering and decreasing the thermal conductivity $\kappa$. But this method has a fundamental lower limit

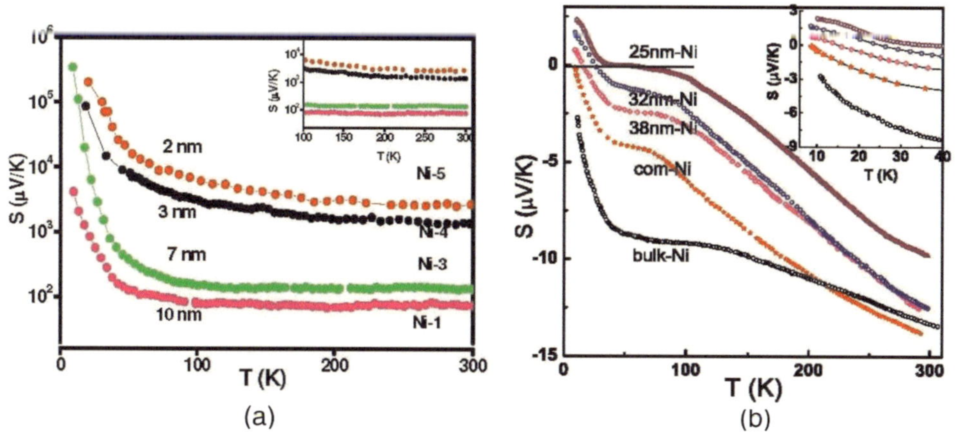


**Figure 7.**
*(a) Variation of thermopower (S) with temperature of compacted Ni-NPs samples for various TOP concentration [34]. Inset shows variation of S in the temperature range 100–300 K; (b) Thermopower of various Ni samples, as indicated [35]. Inset: Low temperature regime.*

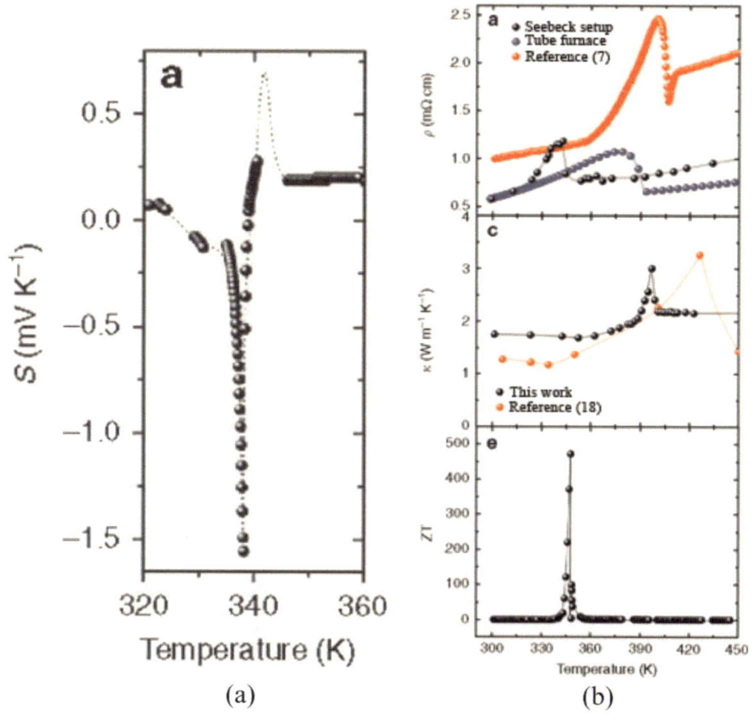


**Figure 8.**
*Temperature dependence of thermopower S, electrical conductivity ρ, thermal conductivity κ, and thermoelectric figure of merit ZT for $Cu_2Se$ around the structure transition [36].*

of about 0.1 W $m^{-1}$ $K^{-1}$, inherent in amorphous materials. Lower value of $\kappa$ in experiments is about 0.2–0.5 W $m^{-1}$ $K^{-1}$ for thermoelectric materials, so one can say that this method almost achieved the limit.

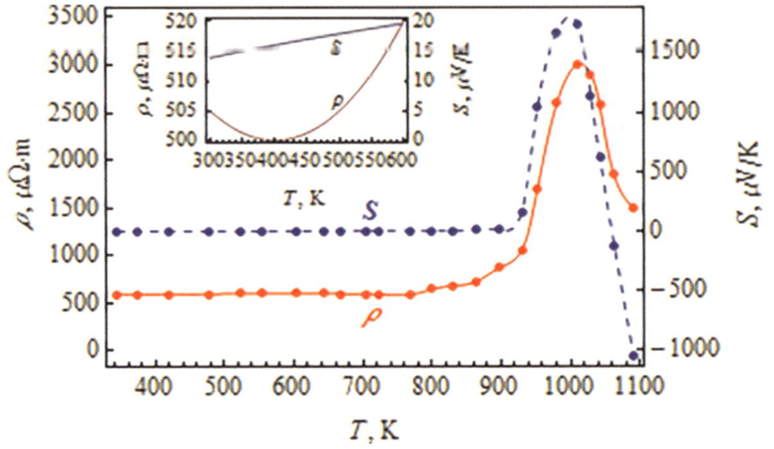


**Figure 9.**
*Temperature dependence of resistivity and Seebeck coefficient for silicate glass doped with $RuO_2$ (thick film resistor). Insert is same about room temperature [25].*

Effect of carrier concentration on $S$ was directly conformed in experiments on thermoelectric profiling of $p$-$n$-junctions, made of GaAs (**Figure 11**) [46] and Si (**Figure 12**) [47]. In these experiments, the heated tip of the scanning tunnel microscope was in contact with the sample along a line perpendicular to $p$-$n$-junction (inset in **Figure 11b**) and the thermoEMF between the tip and the sample was measured at each point. Despite the nanometer resolution, these results do not solve the problem of location of the thermoEMF because of measuring the thermoEMF between the material of heated tip and the cool sample at the point of contact.

## 4. Discussion of theoretical and experimental results

It was established in two center of investigations that thermoelectric phenomena are strong nonlinear for all materials (metals, alloys, and semiconductors), and these nonlinearities manifest mainly at low temperatures (**Figures 3–9**), so structure investigations of samples simultaneously with measuring of $S(T)$ is complicated. Moreover, there are only few experimental data on simultaneously measured $S(T)$ and $R(T)$ for the same samples, especially for metals and alloys. As consequence, the features of $S(T)$ in **Figures 3–9** are interpreted mainly in itself. This situation is some corrected in last years [16, 17, 19].

The Mott formulas (11) and (12), which relate $S(T)$ and $\sigma(T)$ through the density of states $N(E)$, are of little use in the interpretation of experimental data due to the fact that there are often no data on $N(E)$.

Let us now to consider the problem of Seebek coefficient measurement. Direct measurement of $S(T)$ in a homogeneous material is essentially impossible (Magnus law, 1851, [4]), and it is determined from measured $\tau(T)$ via the expression (4) [10, 48]:

$$S(T) = \int_0^T \frac{\tau(T)}{T} dT = S(T) - S(0). \tag{14}$$

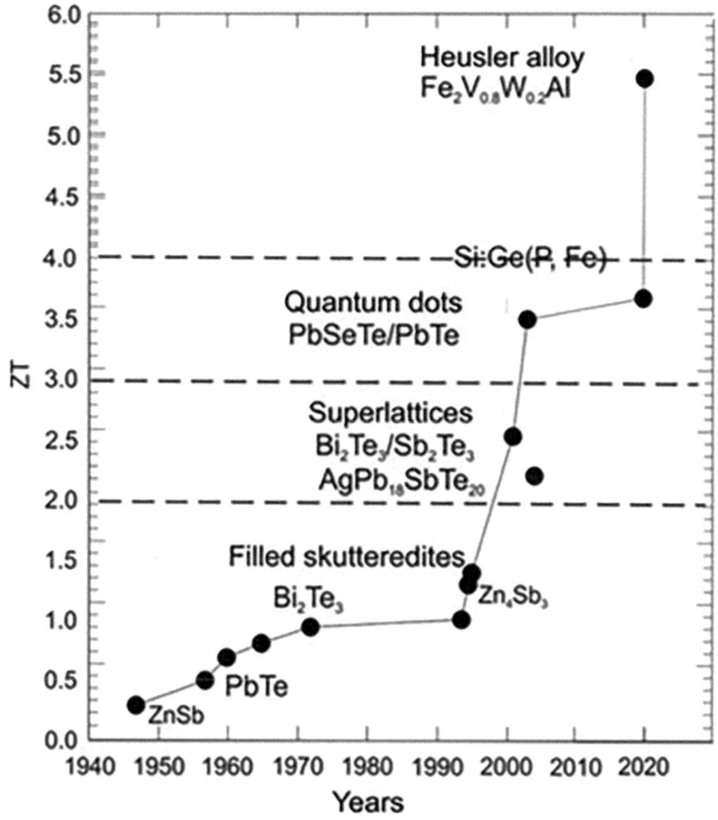


**Figure 10.**
*Achievements in thermoelectric materials. $Fe_2V_{0.8}W_{0.2}Al$ from [37], Si:Ge(P,Fe) from [38].*

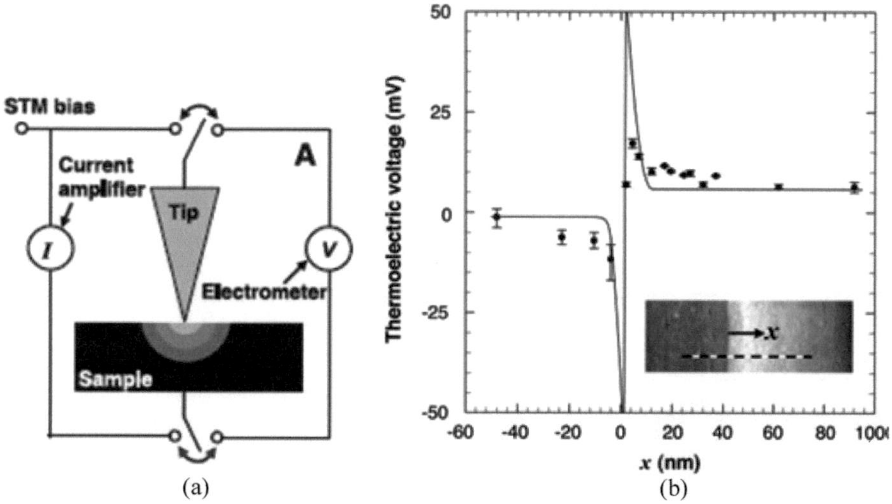


**Figure 11.**
*(a) Schematic of the SThEM setup showing a nonuniform temperature zone in the sample; (b) measured thermoelectric voltage (circles) and calculated S multiplied by 22 K (line) across a GaAs p-n junction as a function of distance (x) [46].*

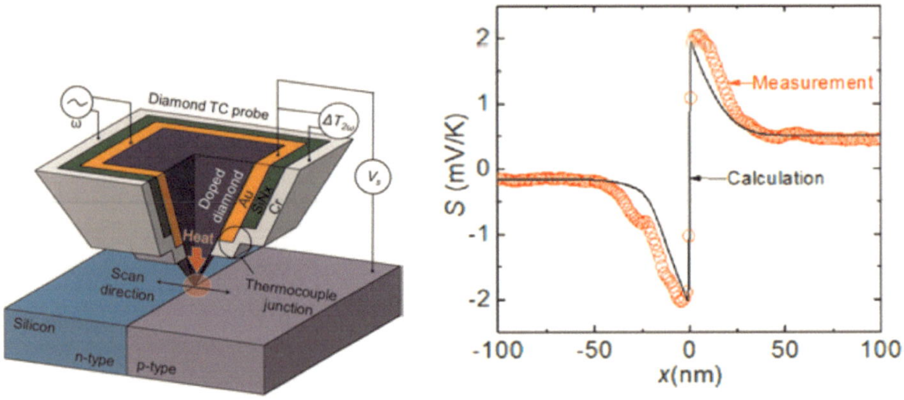


(a) (b)

**Figure 12.**
*(a) Principle of scanning Seebeck microscopy; (b) quantitative profiling of* S *across a silicon* p − n *junction with STM [47]. The theoretical* S *profile based on the doping density and the* S *profile reconstructed from* $V_S$ *and* ΔT.

This situation gives rise to the following problem. In (14), $\tau(T)$ and temperature should be measured from $T = 0$ K, but the third law of thermodynamics forbids reaching 0 K. To get around this obstacle, the assumption is made that $\lim\limits_{x\to\infty} S(T) = 0$. Unfortunately, this assumption is not justified physically in any way. The third law of thermodynamics does prescribe that $\Delta(T = 0) = 0$ and $\lim\limits_{T\to 0} \Delta s = 0$ (the Nernst theorem, $s$ is an entropy). However, (i) the Nernst theorem is applicable only to equilibrium states, while thermoelectric phenomena are essentially nonequilibrium (there are mass, charge, and energy transport, simultaneously); (ii) the definition of the Seebeck coefficient $S(T) = \Delta\varphi(T)/\Delta T$ mathematically leads to an uncertainty of 0/0 at $T \to 0$ K, and the value of $S(0)$ may be quite different depending on the behavior of $\Delta(T)$ near 0 K. In metals and alloys, indeed $S(0) = 0$, but recently, materials have been found [11, 22, 23, 26] in which there is no indication that the assumption $S(0) = 0$ is valid (**Figure 3h, i, 4** and **8**).

This conclusion is additionally confirmed by limit of $S(T)$ from expressions (7) and (9):

$$\lim_{T\to 0} S(T) = \begin{cases} 0 & \text{for metals,} \\ \infty & \text{for non-degenerate semiconductors.} \end{cases} \tag{15}$$

As is noted above (page 2), definition (1) implies by default that $S$ does not depend on temperature (or weakly dependence), so $\Delta T << T$ is required in experiment to provide applicability of (1); that is, $\Delta T$ should be about few K. But then, average excess of phonon's energy $k_B\Delta T \approx 100$ μeV is essentially less than the excess Coulomb energy $e\Delta\varphi \approx 0.4 - 10$ eV of electrons, and it is required to find out a source of their additional energy.

To avoid the difficulties associated with the Seebeck coefficient for the second branch of the thermocouple while measuring the absolute Seebeck coefficient of a conductor, it is advisable to use a superconductor in the superconducting state as the second (reference) branch [27]. The Seebeck coefficient of a superconductor is, by definition, zero. However, this raises the question of the contact of the superconductor with the conductor under study, which is not yet considered.

There is a contradiction between the definition of the Seebeck coefficient (1) and the first Thomson formula (4) as well. Namely, definition (1) requires that $S$ does not depend on $T$, that is, $dS/dT = 0$. So, the Thomson coefficient $\tau = TdS/dT = 0$. This means absence of the Thomson phenomenon in *any materials*.

A similar conclusion follows from the fact due to the assumption of small $\Delta T$; definition (1) can be considered as an expansion of the unknown function $\Delta\varphi(\Delta T)$ in a series in powers of $T$ near $T_0$ (compare with definitions of temperature coefficients of linear expansion, electric resistance, capacities, inductance):
$\Delta\varphi(\Delta T) = (T - T_0)\,(d\Delta\varphi/dT)_{T_0}$, and $S = (d\Delta\varphi/dT)_{T_0}$ must also be constant, so $\tau = TdS/dT = 0$ is again in accordance with (4).

As seen from above, understanding the physical nature of thermoelectric phenomena is not far from the situation at the end of nineteenth century [6, 49] and in the 1950s [3, 5]:

"The subject, I have chosen is one intimately connected with the names of at least two well-known members of this University—the late Prof. Cumming and Sir William Thomson. It possesses at present peculiar interest for the physicist; for, though a great many general facts and laws connected with it are already experimentally, or otherwise, secured to science—the pioneers have done little more than map the rough outlines of some of the more prominent features of a comparatively new and almost unexplored region. Some of its experimental problems are extremely simple, others seem at present to present all but insuperable difficulties. And it does not appear that any further application of mathematical analysis can be safely, or at least usefully, made until some doubtful points are cleared up experimentally." [49].

"The best currently used $Bi_2Te_3$-$Sb_2Te_3$ system was optimized in 1949 ($Z$ increased from $\approx 10^{-3}$ $K^{-1}$ to $\approx 2.6{\bullet}10^{-3}$-$K^{-1}$). Over the following 50 years this system was reoptimized only insignificantly (by no more than 15%) ... By now the method of solid solutions has exhausted itself. Numerous attempts to introduce third and subsequent systems gave only a 2–3% improvement.

Currently, the efficiency of thermoelectric materials (low-temperature) is $Z \approx 3{\bullet}10^{-3}$ $K^{-1}$. The next qualitative step is needed. Various options and new methods are tried and studied. However, they give a significant addition to the $Z$ value only for materials with a low initial efficiency value. Therefore, again and again we can only approach the magic number $Z = 3{\bullet}10^{-3}$ $K^{-1}$. Is it a law of nature? Rather, no. But it is an experimental fact. For bad materials the new methods give significant improvements, for good ones - small ones, for very good ones - practically none so far.

Unfortunately, this explains the barrier that thermoelectric physics and technology cannot overcome. New fundamental models and approaches are needed." [50]

## 5. Possible ways to solve the problems in thermoelectricity

At present, two directions for finding solutions to existing problems of physics of thermoelectricity can be assumed.

**The first one** consists of the accounting of nonlinear correlations of thermodynamic forces and flows in expressions (6) and (7) by including the highest order terms (at least - the second order):

$$j = -L_{11}\nabla\frac{\mu}{T} + L_{12}\nabla\frac{1}{T} + \frac{1}{2}L_{13}\nabla\frac{\mu}{T}\nabla\frac{1}{T} + \frac{1}{2}L_{14}\left(\nabla\frac{\mu}{T}\right)^2 + \frac{1}{2}L_{15}\left(\nabla\frac{1}{T}\right)^2, \qquad (16)$$

$$q = -L_{21}\nabla\frac{\mu}{T} + L_{22}\nabla\frac{1}{T} + \frac{1}{2}L_{23}\nabla\frac{\mu}{T}\nabla\frac{1}{T} + \frac{1}{2}L_{24}\left(\nabla\frac{\mu}{T}\right)^2 + \frac{1}{2}L_{25}\left(\nabla\frac{1}{T}\right)^2 - \mu j \quad (17)$$

Bakhareva [51] confirms that for coefficients of linear parts of expressions (16) and (17), the Onsager's reciprocity relations are again carried out: $L_{12} = L_{21}$. The nonlinear terms cause bifurcation, turbulence, and a hysteresis, complex chemical processes (for example, Belousov-Zhabotinsky reactions), self-organizing of various systems, including biological (up to emergence of life). These phenomena are considered in Prigogine's fundamental works [52] and Bakhareva's [51] and Maksimov's [53] monographs, mainly discussed with reference to the chemical and the biological processes. Stratanovich [54, 55] has considered the fluctuation-dissipation theorems for thermodynamics and their applications to some chemical and the physical systems, for example, to the Benar's cells formation.

Nonlinear properties in thermoelectricity are considered in a number of works [56–59], mainly in the case of the high gradient of temperature arising in the nanostructures. Cimmelli et al. [56] noted violation of Onsager's reciprocity relations because of nonlinear processes. But more detailed consideration of the problem compels one to admit the impossibility of same gradients. Moreover, the conception of temperature comes some undefined at nanoscale.

There are mathematical obstacles for investigation of charge and heat flows (16) and (17) as well.

**The second** possible direction to eliminate the collected divergences of experiment and the theory of the thermoelectric phenomena, most likely, as supplementing previous one, is based on the wave nature of thermal radiation. As a matter of fact, the existing theories of the thermoelectric phenomena provide information regarding the interaction of the thermal radiation with the thermoelectric material as process of energy exchange only. This idea considered a process of energy transmission from a heat to an electronic gas (the heat engine) of density $n$ and pressure of electronic gas $p = nk_BT$ changes. In the absence of the temperature gradient, the difference of pressure of electron gas in contacting conductors is counterbalanced by the difference of contact potential. Difference of temperatures in diverse conductors redistributes mobile charges and generates thermoEMF [6].

It is interesting to note that modern researchers of thermoelectric phenomena argue on the generation of thermoEMF in the volume of a homogeneous conductor by the temperature gradient (neglecting the presence of a contact separating two different conductors). However, in the absence of a temperature difference, the pressure difference of the electron gas in the conductors (due to different $n$ and $E_F$ in them) is compensated precisely by the contact potential difference; that is, the state of contact of conductors is important for the occurrence of thermoEMF. In this regard, we recall once again that the Seebeck and Peltier phenomena are mutually inverse, and the experiment of E. Lenz in 1838 unambiguously proves that the second takes place precisely in the contact of different conductors (see page 655 in [6]).

It is well-known that a thermoelectric system has spatial inhomogeneity of various scale (fluctuations of composition, contacts, polycrystallinity, etc.), and as a result, the concentration of free charge carriers $n(x)$ is spatial inhomogeneous. The last leads to an uneven distribution of the electric field of incident thermal radiation at nanoscale. Interaction of inhomogeneous high-frequency electric field with free charge carriers generates unidirectional forces (ponderomotive forces) and corresponding flow of the carriers [60].

From this point of view, free charge carriers in conductors are considered as oscillating system (plasma or Langmuir oscillations) with own frequency [61]

$$\omega_{\mathrm{p}} = \sqrt{\frac{ne^2}{m^* \varepsilon\varepsilon_0}}, \tag{18}$$

This oscillating system interacts with an incident thermal radiation (which is a wideband electromagnetic waves, Planck formula). This concept leads to the origin of dispersion relation, resonance, and the nonlinearity. Here, $m^*$ – effective mass of electrons, $\varepsilon_0$ - electric constant, and $\varepsilon$ - dielectric permeability of a material. The spatial inhomogeneity of $n(x)$ leads to variations of $\omega_{\mathrm{p}}$ in the thermoelectric system, in the orientation and magnitude of the ponderomotive force from point to point (i.e. polarity reversal of the generated thermoEMF). If $\omega < \omega_{\mathrm{p}}$, the ponderomotive force draws carriers into the higher field region, while for $\omega > \omega_{\mathrm{p}}$, this force throw out carriers same regions; that is, at the frequency $\omega \simeq \omega_{\mathrm{p}}$, generated direct EMF changes the polarity [58].

Studies of semiconductor diodes of various designs [62, 63] showed that even diodes aimed for operation at microwave (frequency higher than 1 GHz) lose detector properties at about 100–200 MHz but become capable of detecting at 10 GHz (**Figure 12a**).

**Figure 12b** shows the dependence of the detected open-circuit voltage $V_{OC}$ on microwave power P at 10 GHz for diode GA402B where one can see polarity reversal of $V_{OC}$ and hysteresis. An arrow pointing down corresponds to an increase in power during measurements; an arrow pointing up corresponds to its decrease. Origin of the hysteresis is not clear yet.

The sign of detected voltage of serviceable diode (curve 1 in **Figure 12b**) is opposite to the sign of thermoEMF (curve 2), inherent for deliberate breakdown diode. Comparison of the **Figure 12a** and **b** makes us assume that the detected voltage (curve 1 in **Figure 12b**) has an unknown physical nature. Furthest experiments on point-contact, *p-n*-junction, and Schottky diodes [64] demonstrated the correlation of detector properties (**Figure 12b**) with the microwave electric field inhomogeneity, generated by various methods.

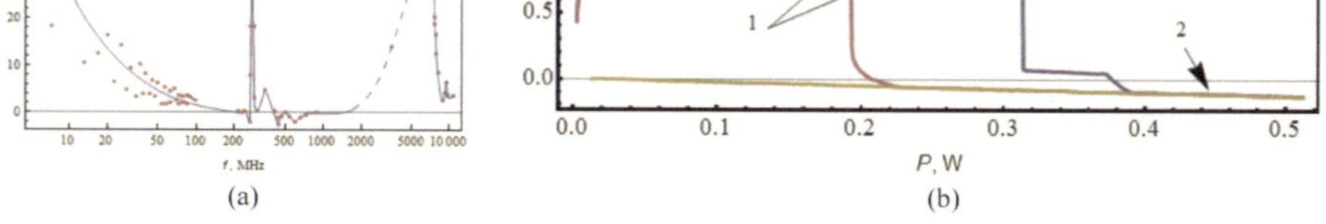


**Figure 13.**
*(a) Frequency dependence of the detected open-circuit voltage* $V_{OC}$ *for the diode GA402B at fixed power P = 1 mW and T = 300 K [62]; (b) the detected open-circuit voltage* $V_{OC}$ *versus power P at 10 GHz for same diode [63]. The germanium diffusion* p-n-*junction mesa-diode GA402B (from Russia) aims to operate as parametric amplifier at 10 GHz [64].*

Probably, observable features of $S(T)$ in various materials (see **Figures 3–9**) are the results of interaction of infrared electromagnetic wave (heat radiation) with the inhomogeneous electron (hole) gas in the thermoelectric and corresponding local variations of $m^*$, $\varepsilon$ or $n$, especially in thin transitional contact region, where strong spatial variations of properties of materials and of the ponderomotive force take place (**Figure 13**).

Such approach to the thermoelectric phenomena enters into consideration, unlike linear nonequilibrium thermodynamics, temporal characteristics of thermoEMF generation process.

## 6. Conclusion

Experiments have confirmed that there are no fundamental limitations for thermoelectric figure of merit $ZT$. Achieved maximum of $ZT \approx 470$ allows us to say that the ideal heat engine has been realized. This engine has efficiency about Carnot's cycle but does not have practical applications because of the narrow working interval of temperature (order of few K).

Origin of the high value of Seebeck coefficient (1000–100,000 μV/K) and nature of their polarity reversal remain unclear yet.

Comparison of the theory of the thermoelectric phenomena and experimental data in various materials (metals and their alloys, oxides, borides, silicides, and intermetallic compounds) shows that observed nonlinear properties (nonlinearity, an extremum, and sign of change) on the temperature dependence of Seebeck coefficient (and, respectively, Peltier and Thomson coefficients) are common for all materials and essentially differ from theoretical representations. It shows that the possible way to eliminate these divergences is the accounting of temperature dependence of parameters of a material (an electrical conductivity, thermal conductivity, and Seebeck, Peltier, and Thomson coefficients) with simultaneous consideration of wave properties of thermal radiation. Also, the comparative experimental study of $S(T)$ and $R(T)$ as well as structural changes on the same sample is necessary.

The correlations between $S$ and $\rho$, on the one hand, and the impurity band position in the energy gap of the silicate glass need thorough investigation in understanding the physical mechanism of electrical conduction and the thermoEMF generation in doped silicate glass, which is considered as prospect thermoelectric material.

From the experimental point of view, it is required to locate the position of the contact where the thermoEMF arises in a homogeneous part of the sample. As far as Seebeck and Peltier phenomena are considered mutually reversible and appearance of the last on contact of two differing materials have unequivocally established, the Seebeck effect needs such experimental demonstration as well.

## Conflict of interests

The authors declare no conflict of interests regarding the publication of this chapter.

G. Abdurakhmanov and G. S. Vokhidova acknowledge the Ministry of Higher Education, Science and Innovation of Uzbekistan for supporting the Uzbek-Indo joint project Uzb-Ind-2021-78 and Uzbek-Belarus joint project IL-4821091667.

D. P. Rai acknowledges the Government of India, Ministry of Science and Technology, Department of Science & Technology (International bilateral Cooperation Division) for supporting the Indo-Uzbek jont project *via* Sanction No. INT/UZBEK/P-02.

## Author details

Gulmurza Abdurakhmanov[1*], Dibya Prakash Rai[2] and Gulbahor Vokhidova[3]

1 The National University of Uzbekistan Physical Faculty, Tashkent, Uzbekistan

2 Department of Physics, Physical Sciences Research Center (PSRC), Pachhunga University College, Mizoram University, Aizawl, India

3 Alfakom Training Center, Tashkent, Uzbekistan

*Address all correspondence to: gulmirzo@mail.ru

## References


[1] Slack GA. New materials and performance limits for thermoelectric coolers. In: Rowe DM, editor. Chapter 34. CRC Handbook of Thermoelectrics. Boca Raton: CRC Press; 1995

[2] Terasaki I. Large thermoelectric power in $NaCo_2O_4$ single crystals. Physics Review. 1997;**B56**:R12685. DOI: 10.1103/PhysRevB.56.R12685

[3] Samoylovich AG, Korenblit LL. Modern state of the theory of thermoelectric and thermomagnetic phenomena in semiconductors. 1. Thermodynamic theory. UFN (Physics-Uspekhi). 1953;**49**(2):243 (In Russian)

[4] Seeger K. Semiconductor Physics. Wien-New York: Springer-Verlag; 2004. 548 p

[5] Samoylovich AG, Korenblit LL. Modern state of the theory of thermoelectric and thermomagnetic phenomena in semiconductors. 2. Kinetic theory. UFN (Physics-Uspekhi). 1953;**49**(3):337 (In Russian)

[6] Khvolson OD. Course of Physics. Vol. 4-2. Petrograd: Rikker's Publishing House; 1915. 1042 p. (In Russian)

[7] Thomson WK. On a mechanical theory of thermo-electric currents; experimental researches in thermo-electricity. In: Mathematical and Physical Papers. Vol. 1. London: Cambridge University Press; 1882. pp. 316-325, 460-470, 654

[8] Onsager L. Reciprocal relations in irreversible processes. I. Physical Review. 1931;**37**:405-426

[9] Onsager L. Reciprocal relations in irreversible processes. II. Physical Review. 1931;**38**:2265-2279

[10] Goldsmid HJ. Introduction to Thermoelectricity. Berlin-Heidelberg: Springer; 2010. 250 p

[11] Ignatov MI, Bogach AV, Demishev SV, Glushkov VV, Levchenko AV, Paderno YB, et al. Anomalous charge transport in $CeB_6$. Journal of Solid State Chemistry. 2006; **179**:2805-2808. DOI: 10.1016/j.jssc.2006.01.016

[12] Ignatov MI. Thermoemf of rare-earth compounds with strong electronic correlations [PhD thesis]. Moscow, Institute of General Physics of the Russian Academy of Sciences. 2006 (in Russian)

[13] Ivanov YV, Vedernikov MV, Ravich YI. Effect of electron-phonon interaction on the thermoelectric properties of superlattices. JETP Letters. 1999;**69**(4):317. DOI: 10.1134/1.568030

[14] Herring C. The role of low-frequency phonons in thermoelectricity and thermal conduction. In: Schön M, Welker H, editors. Halbleiter Und Phosphore/ Semiconductors and Phosphors/ SemiconducteursetPhosphores. Wiesbaden: Vieweg+TeubnerVerlag; 1958. pp. 184-235. DOI: 10.1007/978-3-663-02557-3_12

[15] Gurevich YG, Mashkevich O. The electron-phonon drag and transport phenomena in semiconductors. Physics Reports. 1989;**181**:327-394

[16] Dugdale JS. The Electrical Properties of Metals and Alloys. London: Edward Arnold; 1977. 292 p

[17] Christian JW, Jan J-P, Pearson WB, Templeton IM. Thermo-Electricity at Low Temperatures. VI. A redetermination of the absolute scale of

thermo-electric power of lead. Proceedings of the Royal Society of London. Series A, Mathematical and Physical Sciences. 1958;**245**(1241): 213-221. DOI: 10.1098/rspa.1958.0078

[18] Geballe TH, Hull GW. Seebeck effect in germanium. Physics Review. 1954;**94**: 1134-1140

[19] Geballe TH, Hull GW. Seebeck effect in silicon. Physics Review. 1955;**98**: 940947

[20] Gudayev OA, Malinovsky VK. Temperature dependence of thermopower in polar noncrystalline materials. Physics of the Solid State. 2002;**44**(12):2219-2223

[21] Bentien A, Johnsen S, Madsen GKH, Iversen BB, Steglich F. Colossal seebeck coefficient in strongly correlated semiconductor $FeSb_2$. EuroPhysics Letters. 2007;**80**:17008 (5 p). DOI: 10.1209/0295-5075/80/17008

[22] Littleton RT IV, Tritt TM, Feger CR, Kolis J, Wilson ML, Marone M. Effect of Ti substitution on the thermoelectric properties of the pentatelluride materials $M_{1-x}Ti_xTe_5$(M=Hf, Zr). Applied Physics Letters. 1998;**72**(16):2056-2058

[23] MachidaYo LX, Woun K, Koichi I, Kamran B. Colossal Seebeck coefficient of hopping electrons in $(TMTSF)_2PF_6$. Physical Review Letters. 2016;**116**(8): 087003 (5 p)

[24] Mamedov N, Wakita K, Ashida A, Matsui T, Morii K. Super thermoelectric power of one-dimensional $TlInSe_2$. Thin Solid Films. 2006;**499**:275-278. DOI: 10.1016/j.tsf.2005.07.203

[25] Abdurakhmanov G. On the conduction mechanism of silicate glass doped by oxide compounds of ruthenium (thick film resistors). 3. The minimum of temperature dependence of resistivity. World Journal of Condensed Matter Physics. 2014;**4**(3):166-178. DOI: 10.4236/wjcmp.2014.43021

[26] Behnia K, Jaccard D, Flouquet J. On the thermoelectricity of correlated electrons in the zero-temperature limit. Journal of Physics: Condensed Matter. 2004;**16**:5187-5198. DOI: 10.1088/0953-8984/16/28/037

[27] MacDonald DKC. Thermoelectricity: An Introduction to the Principles. Mineola, N.Y: Dover Publications; 2016. 133 p

[28] Ioffe AF. Physics of Semiconductors. London: Infosearch, Ltd; 1960. 282 p

[29] Mott NF, Davis EA. Electron Processes in Non-crystalline Materials. Oxford: Clarendon Press; 1979. 590 p

[30] Spry M. Rethinking Thermoelectric Effects in Seebeck and Peltier Elements: Toward a Unifying Paradigm. Traverse City: Great Trout Press; 2015. 118 p

[31] Li C, Protik NH, Ordejon P, Broido D. Colossal phonon drag enhanced thermopower in lightly doped diamond. Materials today. Physics. 2022; **27**:100740(30 p). DOI: 10.1016/j.mtphys.2022.100740

[32] Protik Nakib H, Chunhua L, Miguel P, David B, Pablo O. The elphbolt ab initio solver for the coupled electron-phonon Boltzmann transport equations. NPJ Computational Materials. 2022;**8**:28 (9 p)

[33] Inyushkin AV, Taldenkov AN, Ozhogin VI, Itoh KM, Haller EE. Isotope effect on the phonon-drag component of the thermoelectric power of germanium. Physics Review. 2003;**B 68**:153203 (4 p)

[34] Singh J, Kaurav N, Okram GS. Size-dependent thermopower of nickel nanoparticles. In: American Institute of Physics Conference Proceedings Series. Vol. 1591. 2014. pp. 1348-1350. DOI: 10.1063/1.4872954

[35] Soni A, Okram GS. Size-dependent thermopower in nanocrystalline nickel. Applied Physics Letters. 2009;**95**:013101 (4 p). DOI: 10.1063/1.3167302

[36] Byeon D, Sobota R, Delime-Codrin K, Choi S, Hirata K, Adachi M, et al. Discovery of colossal Seebeck effect in metallic $Cu_2Se$. Nature Communications. 2019;**10**:72(7 p). DOI: 10.1038/s41467-018-07877-5

[37] Ghodke S, Muthusamy O, Delime-Codrin K, Choi S, Singh S, Byeon D, et al. Distinctive thermoelectric properties of supersaturated Si-Ge-P compounds: Achieving figure of merit *ZT*> 3.6. 2019: 36. Available from: https://arxiv.org/ftp/arxiv/papers/1909/1909.12476.pdf

[38] Hinterleitner B, Knapp I, Poneder M, Shi Y, Müller H, Eguchi G, et al. Thermoelectric performance of a metastable thin-film Heusler alloy. Nature. 2019;**576**:85-90. DOI: 10.1038/s41586-019-1751-9

[39] Xu Ning X, Yong and Zhu Jia. Topological insulators for thermoelectric. npj Quantum Materials. 2017;**2**:51(9 p). DOI: 10.1038/s41535-017-0054-3

[40] Lee M-J, Ji-Hoon A, Ho SJ, Heo H, Gi JS, Woo L, et al. Thermoelectric materials by using two-dimensional materials with negative correlation between electrical and thermal conductivity. Nature Communications. 2016;**7**:12011(7 p). DOI: 10.1038/ncomms12011

[41] Abdurakhmanov G, Abdurakhmanova NG. High temperature anomalies in resistivity and thermoelectric power of thick film resistors. Physica Status Solidi (A), Applied Research. 2005;**202**:1799-1802. DOI: 10.1002/pssa.200420036

[42] Abdurakhmanov G. On the conduction mechanism of silicate glass doped by oxide compounds of ruthenium (thick film resistors). 2. Nanocrystals in the glass and charge Carrier's localization. American. Journal of Materials Science. 2011;**1**(1):12-17. DOI: 10.5923/j.materials.20110101.01

[43] Abdurakhmanov G, Vakhidova GS. Structural transitions of silicate nanocrystals in the glass. American Journal of Materials Science. 2012;**2**(1): 37-40. DOI: 10.5923/j.materials.2012 0201.07

[44] Abdurakhmanov G, Shimanski VI, Onsengendler B, Umirzahov B, Urokov AN. Pseudogap, nanocrystals and electrical conductivity of doped silicate glass. Technical Physics. 2021; **66**(2):269-274. DOI: 10.1134/S106378422102002X

[45] Abdurakhmanov G. Electrical conduction in doped silicate glass (thick film resistors). In: New Insights into Physical Sciences. Vol. 4. London-Hooghly: Book Publishers International; 2020. pp. 47-71. DOI: 10.9734/bpi/nips/v4

[46] Ho-Ki L, Khajetoorians AA, Li S, Pipe Kevin P, Ram Rajeev J, Ali S, et al. Profiling the thermoelectric power of semiconductor junctions with Nanometer resolution. Science. 2004;**303**:816-819. DOI: 10.1126/science.1091600

[47] Lee B, Kim K, Lee S. Kim Jong Hoon, Lim Dae soon, kwon O and Lee Joon Sik. Quantitative Thermopower profiling across a silicon $p-n$ junction with Nanometer resolution. Nano Letters.

2012;**12**:4472-4476. DOI: 10.1021/nl301359c

[48] Kikoin IK, editor. Tables of Physical Quantities. Moscow: Atomizdat; 1977. 1008 p., p. 462 (in Russian)

[49] Tait PG. First approximation to a thermo-electric diagram. In: Scientific Papers. Vol. 1. Cambridge: Cambridge University Press; 1898 [Transactions of the Royal Society of Edinburgh, Vol. XXVU. Read December 1, 1873.]

[50] Iordanishvili EK. Introduction to thermoelectricity. Thermoelectric effects. In: Bulat LP, editor. Thermoelectric Cooling. St-Petersburg: St-Petersburg University of Low-temperature and Food Technologies; 2002. 146 p. P. 9-38 (In Russian)

[51] Bakhareva IF. Nonlinear Nonequilibrium Thermodynamics. Saratov: Saratov State University Publishing House; 1976. 141 p. (in Russian)

[52] Kondepudi D, Prigozhin I. Modern Thermodynamics: From Heat Engines to Dissipative Structures. Chichester: John Wiley and Sons; 2015. 525 p

[53] Maksimov AI. Introduction to Nonlinear Physical Chemistry. Ivanovo: Ivanovo State University of Chemical Technology; 2010. 175 p (in Russian)

[54] Stratonovich RL. Nonlinear Nonequilibrium Thermodynamics I. Linear and Nonlinear Fluctuation-Dissipation Theorems. Berlin, Heydelberg: Springer; 1992. 377 p

[55] Stratonovich RL. Nonlinear Nonequilibrium Thermodynamics II. Advanced Theory. Berlin, Heydelberg: Springer; 1994. 240 p

[56] Cimmelli VA, Sellitto A, Jou D. A nonlinear thermodynamic model for a breakdown of the Onsager symmetry and the efficiency of thermoelectric conversion in nanowires. Proceedings of the Royal Society A: Mathematical, Physical and Engineering Sciences. 2014; **A470**:2170(13 p). DOI: 10.1098/rspa.2014.0265

[57] Zhang T. Analytical solution of nonlinear thermoelectric heat transport equation using Homotopy perturbation method. Journal of Computational Intelligence and Electronic Systems. 2015;**4**(1):59-66. DOI: 10.1166/jcies.2015.1115

[58] Yamashita O. Effect of linear and non-linear components in the temperature dependences of thermoelectric properties on the energy conversion efficiency. Applied Energy. 2009;**86**:1746-1756. DOI: 10.1016/j.enconman.2009.04.019

[59] Benenti G, Casati G, Prosen T, Saito K. Colloquium: Fundamental aspects of steady state heat to work conversion. 2013. arXiv:1311.4430v1 [cond-mat.stat-mech]

[60] Chen FF. Introduction to Plasma Physics and Controlled Fusion. 3rd ed. Heidelberg: Springer; 2016. 496 p

[61] Artsimovich LA, Sagdeev RZ. Plasma Physics for Physicists. Moscow: Atomizdat; 1979. 315 p (In Russian)

[62] Abdurakhmanov G, Esbergenova A, Reyimbaeva S. Anomalies of frequency cutoff of semiconductor diodes at microwave. Technical Physics Letters. 2021;**47**(1):5-7. DOI: 10.1134/S1063785021010028

[63] Reymbaeva S, Abdurakhmanov G, Orel A. Nonlinear properties of an inhomogeneous diode structure in a strong microwave field. World Journal of

Condensed Matter Physics. 2023;**13**:1-13. DOI: 10.4236/wjcmp.2023.131001

[64] Microwave detector diodes, technical data and parameters (In Russian). Available from: https://eandc.ru/news/detail.php?ID=18813 [Accessed: 2023-06-28]